CELEBRATING THE SEASONS WITH

The

Yorkshire Shepherdess

Amanda Owen

MACMILLAN

There's no mountain that you cannot climb.

Contents

Introduction

Am I a photographer of critical acclaim? No. A writer of literary brilliance? No. I am a shepherdess, mother, wonderer and dreamer. As a child I was always a reader, though never a writer. What mattered to me was finding authenticity in all that I read. I wiled away countless hours poring over books, but it was one particular volume that really spoke to me – *Hill Shepherd* by John and Eliza Forder, a photographic journal that followed the lives of hill shepherds in the Yorkshire Dales and Lake District. Although I was living in Yorkshire, in Huddersfield, I felt a world away from the mountains, moors and fells that filled the pages. The proximity to nature, the connection with the animals and the wild freedom that came with shepherding sheep fascinated me, and from that moment onwards it became my sole ambition and goal in life. I established myself as a contract shepherdess, slowly working my way up the farming ladder, gaining practical know-how and knowledge in the complicated and often-closed world of shepherding. By twenty-one I was married to Clive Owen and was living at Ravenseat, one of the highest and most remote hill farms in England.

I started sharing the story of life on the farm through posting pictures on social media, writing books and later appearing on television. I feel privileged to live and work at such an outstandingly stunning place and enjoy documenting the

ever-changing seasons through my photography, come rain or shine and everything in-between – as is often the case with Yorkshire's weather!

I don't see myself as a photographer, or even an author for that matter; just a visual diarist and an observer of what goes on around me. Every which way you look there is a photograph to be had, a combination of children and animals – the two things you are told to never work with – set against the glorious backdrop of Yorkshire's big skies and dramatic landscapes. The spontaneous nature of photos that aren't choreographed are what I seek to share. I prefer to turn away from the classic view, to find something unconstrained and natural, simplicity itself, a snapshot of a moment in time.

I am very aware of how fortunate I am to be in a position to share some of what our great British countryside has to offer, whether in the practical sense of the food we put on our plates or, from the more holistic point of view, the mindfulness and healing powers that nature can bring into our busy, frantic lives.

The recipes you will find in this book are not meant to be followed to the letter, but are merely starting points. Switch the vegetables according to what you have in your cupboard, substitute and

adapt whenever and wherever you can, make use of what you already have available – whatever will make your life simpler.

This is not intended to be an instruction manual for how a life should be lived; I am the last person who would see my life as being a perfect example for others. This is simply an account of life at Ravenseat, a glimpse into the place where we live and work, the family dynamics and our interactions with nature and the countryside within the specific confines of our farm. It follows what is happening outdoors and indoors through the seasons, and the link between the two.

I hope you take the same pleasure from *Celebrating the Seasons* that I found in immersing myself in the books of my youth, when life as a shepherdess was only a dream . . .

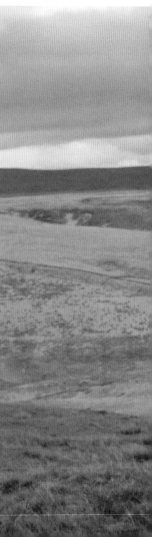

January

I'd challenge anyone to find a more dramatically spectacular place to be in January than a snowbound Ravenseat. Granted, a kinder, simpler home would be easier, but for wild dynamic beauty, Ravenseat has to win, hands down. The farmstead sits neatly and squarely in a small valley, cosily hemmed in and dominated by the moors that surround her. With the farm comes a well-documented history going back centuries, the human presence visible in the walls and buildings. Much older, still, are the fossils still visible in the sandstone cobbles dug from the nearby quarry, and the water-worn swirls on flagstones hauled from the beck.

It is pretty much a certainty that all will be covered in a thick blanket of snow, drifts gathering beside the walls, delicate and pure windblown sculptures that glisten in the rare moments that a weak winter sunshine breaks through the heavy leaden skies so common at this time of year. Blizzarding snow and ice mean that we may well be cut off for at least part of the month, but we know the drill by now, and make sure we are well prepared. I fill the dairy with staples – sacks of rice, potatoes and pasta, big bags of sugar and flour and plenty of

A fully stocked dairy is a must for the winter months.

tins. It is well within the power of every farmer to furnish the table through all seasons with wholesome dishes from the dairy, if preparations have been made.

I drive thirty miles to Catterick Garrison where I can buy in bulk, partly in a bid to maximize efficiency by keeping trips to a minimum, and partly as it makes financial sense. With a family to feed, and a sizeable one at that – eleven people round the table, all with healthy appetites – keeping hunger pangs at bay is no easy feat in winter. Normally bulk-buying poses no issue, but in 2020, when the pandemic lockdown was introduced, it became far more difficult to stock up in the way in which I was accustomed. Furtively loading up a trolley, I would constantly be looking over my shoulder, aware that to my fellow shoppers it would seem as if I was greedily hoarding food.

I get potatoes by the sackful from the local agricultural suppliers, who stock them alongside the animal feed. The price varies depending on the season and availability, but even at their costliest, £12 for a twenty-five-kilo sack is still a bargain. They are unwashed, encrusted with a little soil so they store well without rotting or sprouting. They will keep for weeks if left in a cool, dark place. How we store food is so very important – waste is to be avoided at all costs. It is undoubtedly difficult to incorporate much in the way of seasonal fruit into our diets. Other than cooking apples, frozen autumn berries and stewed rhubarb there's little to be had. But that is how it should be; periods of plenty (a glut, if you like) and times of scarcity. An awareness of seasonality is of huge importance, and often overlooked in a world where everything is available at all times. There is a time for everything, and it comes as no surprise that eating what is in season is easier on the pocket, and naturally more nutritious and flavoursome.

While I leave everyone to sort their own food for dinner time, around midday, everyone pulls up a chair and gathers around the kitchen table for tea. Talk will inevitably be about the animals, and the jobs that were done that day. We make vague plans for the following day, working out logistics to maximize efficiency and tackle problems. Our enjoyment is as much about the conversation as it is the meal itself, but it is during these coldest of days that everyone craves filling, hearty meals. 'It's yer belly that keeps yer back up', as they say, so we will eat potatoes all ways, heaps of root vegetables, roasts, soups with homemade bread, and pies, sweet and savoury, crumbles and custard.

Once the plates are sided away it is back outside to do the bullocking up – the term used for feeding up and bedding all of the housed animals and readying them for the night. Everyone has their own chores to do, from feeding the dogs and hens to filling up the log basket and coal buckets. It's all to do before nightfall. We milk our house cow twice a day, first thing in the morning and last thing at night; twelve hours between milking is ideal so there's no undue pressure on the udder. And cows are creatures of habit who get to know

and love regular routines. It settles them if they know what to expect, and when.

The three smallest children, Clemmy, Annas and Nancy, like to sort out the stables. They are very capable of stuffing haynets and poo-picking. Sidney and Violet see to the calves, mixing up milk bottles or buckets and filling their hay-racks. Miles often lets the sheepdogs out of their kennels for a gallop up to the beck, where they can get a drink. If they have worked with us all day, they do sometimes choose to stay in the kennel, no doubt tired from all the miles covered.

Edith is tasked with bottling the milk from the house cow. I bring her a pail of warm milk and she will first strain it through a stainless-steel sile with a circular dairy pad that fits in the bottom, then decant it into glass screw-top bottles. Sometimes we change tactic, putting the milk into a wide dairy bowl and letting it cool in the fridge. Later we skim off the cream to use in ice cream or in a pudding. After Raven left for university, Edith was determined to take on more responsibility,

and now deals with a good many of the indoor chores, too, such as helping to clear away dinner and giving me a hand with sorting through the washing.

Miles is still fanatical about his chickens. Reuben, too, never changes, and will stay outside in the tool shed crafting away at some mechanical project until bedtime.

When Raven is back home I rely on her massively. She knows what needs tackling, and gets stuck in, whether it is dealing with the horses in the field, getting the little ones bathed, or whipping up a batch of bread.

I can honestly say that it's rare for me to have to say a wrong word to the children when it comes to their chores. Everyone knows what needs doing and gets on with it. And I am hardly a real home-maker, and can't expect things to be 'just so'. That is not in my nature. Ravenseat is not a show home – there is too much clutter and busyness for it to be orderly – but it is a place that invites, exudes warmth and comfort. Unless you stand on Lego or sit on a dozing terrier – but that's the risk you run in our home.

We raise Swaledales here, a distinctive, hardy, native breed with coarse wool, long tails, mottled legs and black-and-white faces. During January our main task is to keep them fed and safe. We have around 850 of them, and if all has gone to plan the majority of the females will be in lamb by January. Ravenseat comes with grazing rights for the moor, and we leave sheep on their heafs on the moortops for as much of the winter as possible. It is where they belong, and where they are at their happiest.

When the weather is clashy, and the wind is blowing, we often find the sheep 'hurling' – stood with their heads down, backs up and bottoms turned towards the wind, their thick tails providing

protection for their udders. The practice of docking tails is not required here, where the tail is seen as a valuable asset in keeping the weather at bay. A long tail can become soiled and attract flies, but this tends to happen in areas with warmer weather and more tree coverage, which means more flies, including bluebottles, and where flystrike is a concern. Prevention is better than a cure, so in those areas docking will stop a mucky build-up and save the poor sheep from being eaten alive.

Being a shepherdess brings with it a fair share of responsibility, and never more so than in the depths of winter when we will see our sheep every day, familiar faces come looking for their daily rations of food. Our job is to keep our sheep healthy and settled, and we know their welfare rests on our shoulders. We deal with day-to-day minor ailments such as foot rot and blind illness ourselves, but we get the vet out for anything of a more serious nature. We take an annual review of the health status of our flock and herd so that we can address any issues that might have arisen. One year we had a higher-than-average number of abortions in the flock, and by blood-testing we were able to ascertain the strain of virus that was to blame and vaccinate accordingly. The cows are part of an accredited High Herd Health Scheme, which means they are all blood-tested annually by the vet, and we are then given a certificate that states our animals are free from highly transmissible diseases such as BVD (bovine viral diarrhoea), Johne's disease and tuberculosis.

Our youngest three girls are very willing aides during visits by the vet, asking a whole host of questions and offering up advice, whether wanted or not. Recently Annas was proudly wearing the

stethoscope around her neck after being shown how to listen to a calf's heartbeat and the noises in its stomach – a sign of healthy rumination. Once the patient had been released back to the herd the children began to listen to each other's heartbeats. What I hadn't noticed was that the other indicator of good health, the checking of body temperature, had also been undertaken by the vet. Obviously, since this was a bovine, the thermometer had not gone under the tongue . . .

We had only turned our backs for a minute when I glanced over to see Clemmy taking Nancy's temperature, the thermometer in her mouth.

Generally a serenity pervades at the start of the year, a feeling that January is almost the calm before the storm, but not always . . . As I write in 2021 there has been a snowfall every day since Christmas, a dry, light snow, that was picked up and sent swirling around by the wind. Layer upon

layer of powder-fresh snow made a great surface for the children to sledge and snowboard, but wasn't so good for snowballs, which wouldn't stick together without the help of some muck from the midden – not the most pleasant missile to get in the face!

There was no talk of a storm, just a gradual accumulation of snow, day after day. The sheep were coping well, settled on their heafs – the area of moorland they recognize as their own – at the moors, so we let them be, foddering them daily. Then one morning we woke to a heavier snowfall, enough to prevent us from taking a vehicle to the sheep. Clive and I went to the heafs together on foot, an epic undertaking at the best of times; walking uphill, bent double under the weight of bales of hay. It was hard going, and the higher we climbed the deeper the snow became and the more we were forced to exaggerate our steps. Meanwhile

the wind whipped up loose snow which stung our already-reddened faces.

We both whistled up the 175 sheep that were out in this heaf, and from the snow-white wilderness they came, pleased to see us – or at least our offerings.

While they don't have boundaries, the sheep tend to stick to their heaf and certainly don't stray far during hungry times, as they learn to expect their daily ration of food. Also they are not really loners, seeing safety in numbers. We were definitely short on a few that day. Laying out the canches of hay in a long line meant we could count accurately while they had their heads down eating. We had only 155. By now it was stowering – the wind had picked up and visibility was poor, so much so that I couldn't see where the sky ended and the snow began. Hunt-

ing for my missing twenty in these conditions was pointless, and we retired back to the farmhouse, cold and tired, to wait for a break in the weather.

Two days later, it was a bright, crisp, calm morning, so Clive and I headed out to find our missing sheep. There was no point setting off in vehicles. We attempted to go part of the way on the tractor but the moment we hit a drift it slew sideways, tipping awkwardly, wheels spinning. We were fortunate we did not roll her right over, such was the precarious position we found ourselves in. It was way too deep for any thoughts of riding to the rescue on horseback. The dogs, too, had to stay behind, as with every step in snow like this, they disappear above their head.

It was up to our thighs at points and we were forced to crawl, with no breath to speak. We would

pause, whistle, shout and listen, straining our eyes, but there was no sign of movement, only white, and more white. With such a large area to cover, we were clueless about which direction to walk in.

Every day for a week, we went through the same routine, with no luck, and we were beginning to despair. It was such light, blowy snow that I didn't think the sheep had been buried under it, and they can actually survive in it a long time provided they don't get trapped upside down. I had to hope they had found shelter, but I couldn't settle until I knew they were safe.

'You need a skidoo,' Reuben mentioned thoughtfully. 'I've got a mate wi' yan in his garage.'

I can't say I was really convinced that Reuben had a friend with a skidoo but he quickly proved otherwise when, a couple of hours later, I heard what can only be described as a jet engine in close proximity. Hurtling towards the farm was a bright yellow skidoo. Martin, Reuben's friend, had driven it the fifteen miles from his home in Kirkby Stephen and was sat triumphantly astride the machine, gloriously resplendent in a military snowsuit, helmet and goggles, with a cigarette hanging from his mouth at a rakish angle.

Before I knew it, Reuben and Miles had hopped on the back, and they were off. Hours later the sky was darkening, as were our thoughts. 'We'll be out looking for 'em an' all, as well as t' sheep,' Clive muttered, shaking his head. But he was wrong, for already we could hear the sound of an engine approaching and, sure enough, they reappeared, jubilant.

'We've found five of yer sheep,' Martin announced.

ABOVE **At Graining Scars Seat with Kate.**
LEFT **Martin's skidoo!**

There are outlying stone folds dotted all over the moor, an ancestral throwback to a time when the shepherds stayed out with their sheep, overnighting in simple shelters, some of which are still intact, the all-important fireplaces visible, too. It must have been a lonely and isolated existence, sleeping and waking with the daylight, and living off oatcakes, cheese and dried meats. It was a way of life for many until the late nineteenth century, when things began changing, but until then, much of the shepherding was done in situ out on the moors, using the outlying pens to do the work, shearing the sheep with hand-clippers and salving them. In one of these folds, five of our sheep had gained respite from the wintery conditions. There was no possibility of moving them, as they would start to sink in the deep snow and a sheep's reaction to that situation is to dig in and refuse to go further. All we could do for the moment was take them hay and await a bit of a thaw.

After loading the skidoo with hay, I was given a quick lesson in riding it. The simple concept is that the skidoo travels on top of the snow's surface – and the faster you go, the smoother the ride, with a top speed of 70 mph! It was terrifying. But with a new tool in our repertoire, we carried on searching and found two more of the sheep in a hilltop pen on the other moor; they must have crossed the river and been blown there by the storm. It was a stressful ten days before we found the remaining wanderers, right at the top of the valley. They were fine, other than the fact that, as a result of their newfound freedom, four of them had gone wild. It was impossible to get close to them, and even the sheepdogs couldn't round them up. It is hard to explain, but within a very short time of looking after themselves, the tamest sheep can turn wild. There was nothing for it but to leave those four up there, knowing that when the weather improved we would be able to bring more people and dogs.

At the same time we had another big problem – no running water in the farmhouse. I've got a weather station in the farmyard, and could see that the temperature had dropped to -12°C during the night. Our water system is piped down from a natural spring further up the moor, and we worried that the water in the pipes would freeze.

'We'd better put the tap on to drip, just to keep it flowing,' I said to Clive, then we went to bed.

At 4 a.m. we were woken by the cows bellowing loudly. Clive swiftly went outside only to discover that their trough was dry. Cows drink copious amounts of water, especially when they are producing milk for suckling calves, so he spent a good hour stomping around in the dark, trying to work out exactly where the problem lay. We are fortunate to have a supply of our own spring water, but when it ceases flowing it is our responsibility to rectify any issues.

Once the children were up, we relayed the bad news, knowing from a previous experience that this would be a pretty miserable period. Everything that normally involved water had to happen in the river. Buckets of water were carried to the house to flush the toilets and for the washing-up, but after breaking the ice beside the bridge we could all clean our teeth in the river!

As for our thirsty cows, we were bucket-filling their water trough, but they were insatiable and would slurp it down without breaking stroke. So we let them out, planning to walk them along to the little trickling beck in the snow, downstream from where we took our ablutions to avoid any contamination. I was not concerned about them getting a hint of milky toothpaste, but would find it wholly unacceptable to have a cow pat in my

We found two more of the sheep in a hilltop pen on the other moor; they must have crossed the river and been blown there by the storm.

teeth-rinsing water. We expected them to drink to their hearts' content then amble back into the barn. All did not go to plan. Not quite. They took one look at the snow and their freedom, put their tails in the air, and went on the rampage, running and jumping around like children let loose in the snow for the first time. They were rolling, and wrestling, their udders swinging wildly, as they had the time of their lives – before returning to the barn, hot, sweating and steaming, not having made use of the river, and now thirstier than ever.

The problem was a broken pipe near the spring, and we were lucky this time that it was fixed in a matter of days rather than weeks. It was a reminder that we should never take for granted our ability to turn on a tap and have a glass of cool, clean and clear water.

We had to leave some of our missing sheep in the fold until we could safely move them.

Hearty Barley and Vegetable Soup

People often question if soup can make for a filling meal – well this one certainly can! I get a big pot of this on the go, and everyone dips in for large bowlfuls.

Prep time 15 minutes / Cooking time 1 hour 10 minutes / Serves 4

INGREDIENTS

30g butter

1 tbsp Yorkshire rapeseed oil

300g carrots, peeled and
   chopped

1 stick celery, chopped

300g potatoes, peeled and
   diced

300g turnips, peeled and
   diced

150g leeks, sliced

250g parsnips, peeled and
   chopped

100g pearl barley, rinsed and
   drained

85g split red lentils, rinsed and
   drained

3 tbsp tomato puree

1.3 litres vegetable stock

2 tsp dried thyme

2 bay leaves

75g kale, chopped

1 small bunch parsley, finely
   chopped

salt and ground black pepper

METHOD

1. Melt the butter and oil in a large pan over a gentle heat. Add the carrots, celery, potatoes, turnips, leeks and parsnips and fry for 10 minutes until the vegetables soften.

2. Stir in the pearl barley, lentils, tomato puree and vegetable stock, then sprinkle over the dried thyme and bay leaves. Bring the soup gently to the boil, then cover and simmer over a low heat for 1 hour until the barley is tender.

3. When the soup has 10 minutes left to cook, stir in the kale and simmer gently until the kale has softened. Season to taste, then stir through half of the chopped parsley.

4. Serve the soup in warm bowls with a sprinkling of the remaining parsley and a chunk of herby focaccia bread (see page 33).

Tip
As an alternative, try adding some cooked leftover chicken and use chicken stock instead of vegetable stock.

Herby Garlic Focaccia

Focaccia bread sounds fancy but it is very simple to make and just the addition of salt and herbs really brings it up a level.

Prep time 15 minutes, plus 1 hour 30 minutes to prove / Cooking time 25 minutes / Serves 4

INGREDIENTS

500g strong bread flour, plus
    extra for kneading
    and rolling

7g fast-acting dried yeast

1 tsp caster sugar

1 tsp fine salt

3 sprigs rosemary, leaves finely
    chopped

1 small bunch thyme, leaves
    finely chopped

1 small bunch parsley, finely
    chopped

350ml warm water

4 tbsp olive oil, plus extra for
    greasing

2 cloves garlic, crushed

2 tsp flaky sea salt

METHOD

1. Put the flour into a bowl and stir in the dried yeast, followed by the sugar and fine salt, then stir in half of the finely chopped herbs.

2. Measure the warm water in a jug and stir in 2 tablespoons of the oil. Make a well in the centre of the flour and stir in the liquid, adding it gradually until you have a slightly sticky dough.

3. Turn the dough out onto a lightly floured surface and knead for 8–10 minutes until the dough is smooth. Transfer the dough to a lightly oiled bowl, cover with a damp tea towel or oiled cling film and leave to prove in a warm place for about an hour until the dough has doubled in size.

4. Lightly grease a 22 x 30 x 2cm tin with oil and gently roll out the dough on a lightly floured surface to fit the tin. Mix the remaining oil with the garlic then drizzle over the dough. Cover again with a damp tea towel or oiled cling film and leave in a warm place for about 30 minutes, until doubled in size.

5. Preheat the oven to 220°C/200°C fan/gas 7. Press your fingers into the dough to make dimples across the surface. Sprinkle over the sea salt and the remaining chopped herbs.

6. Bake the focaccia for 20–25 minutes until the bread is well risen and golden. Allow the focaccia to cool slightly, then turn out onto a board and cut into squares.

Tip
If you prefer, use an electric mixer with a dough hook and reduce the kneading time to 5 minutes on a medium speed.

Clementine Upside-down Sponge

There are inevitably clementines left over from Christmas, sitting in the fruit bowl going wrinkly, and this is a great way to use them up. After the heavy excesses of Christmas, a sponge makes a nice light change.

Prep time 20 minutes / Cooking time 40 minutes / Serves 4–6

INGREDIENTS

Clementine syrup

3 clementines

75g golden caster sugar

1 star anise

1 cinnamon stick

Sponge

170g unsalted butter,
   softened, plus extra for
   greasing

75g golden syrup, warmed

170g golden caster sugar

3 medium eggs, lightly beaten

140g self-raising flour

30g ground almonds

½ tsp baking powder

Vanilla custard

400ml whole milk

200ml double cream

1 vanilla pod, sliced

4 medium egg yolks

50g golden caster sugar

2 tbsp cornflour

METHOD

1. To prepare the syrup, thinly slice the clementines with the peel on. Bring 125ml water to the boil in a medium-sized pan and add the sugar, stirring until it dissolves. Add the star anise and cinnamon stick to the pan, then bring the syrup to the boil. Place the clementines into the syrup, then lower the heat and simmer gently for 5 minutes. Using a slotted spoon, remove the clementines from the pan to a plate and discard the syrup.

2. Preheat the oven to 180°C/160°C fan/gas 4. Line a 21 x 6cm round cake tin with baking parchment and lightly grease with butter. Pour the golden syrup evenly into the base of the tin, then arrange the drained clementine slices on top.

3. For the sponge, place the butter and sugar into a mixing bowl and whisk for 2 minutes until light and fluffy, using an electric whisk. Add the eggs with 2 tablespoons of the flour and whisk for a minute until combined, then add the remaining flour, almonds and baking powder and stir gently until combined.

4. Spread the cake mixture evenly on top of the clementines, then bake on the middle shelf of the oven for 40 minutes, until the cake is golden, firm to touch, and when a skewer is inserted it comes out clean. Allow to cool slightly in the tin, then turn out onto a plate.

5. Meanwhile, while the cake is cooking make the custard. Put the milk, cream and vanilla into a medium-sized saucepan and gently bring to the boil. Whisk the egg yolks, sugar and cornflour in a large jug, then gradually pour the hot milk mixture over, whisking constantly.

6. Remove and discard the vanilla pod, then pour the custard through a sieve into a clean pan. Gently heat the custard, stirring with a wooden spoon until it thickens.

7. Serve the clementine upside-down sponge in bowls with the warm custard.

Tip

For variety, try a peach or plum upside-down sponge. Use the leftover egg whites to make meringues.

February

Elsewhere in the country snowdrops might be starting to peek through, signalling the end of winter, but there's no chance of that at Ravenseat, with the cold and snow set to dominate for a good while yet. Snow on the ground as far on as 1 May is not uncommon, in winter's last hurrah. February still feels like the depth of winter, and the weather has a real sting in its tail, with the cold penetrating to the bone, unless you are keeping active. You get a true appreciation of what it is to be warm when you feel the cold on the top of the moors at this time. There is no pause for complaints though, as the yows (female sheep who have given birth) are now well into their pregnancies, and need us to be taking care of them, more than ever.

In these temperatures, water freezes at the higher points of the river, but a sudden warm spell can cause it to thaw quickly, and break off in huge chunks, like icebergs. About five years ago large lumps of ice began gathering up under the bridge, getting stuck and damming it up. The water was filling back up behind it, and we could only watch in astonishment when, all of a sudden, there was a loud sucking sound, and it was all gone. The whole event had taken mere seconds, and I'd almost have thought I'd imagined it, but for the flotsam providing evidence of how high the water had been.

In 2021 I was in the yard when I realized the sound of the river had a different tone, heavier and deeper. Clive and I headed for a look, and to our amazement there were dozens of icebergs making their way down towards us, some upended and craggy, others great flat hunks of ice, as big as a pick-up truck. They were gathering at the shallow point by the ford, and backing up, with the water levels rising around them. I started throwing rocks to try to loosen the main culprit and get the river flowing again, but it didn't shift an inch.

'There's only one thing for it,' Clive announced, 'you're going to have to stand on it.'

'Really?'

'Aye,' he nodded, 'hop on.'

So on I jumped – and fell straight in. The icy water rushed into my wellies and I struggled my way back to land, gasping from the shock of the cold, desperate to empty the biting slush from around my feet.

But at least it had done the trick – Amanda 1, Mother Nature 0 – although fair to say, beating the weather is a very rare occurrence . . .

The weather let us know just how much it was boss in 2020 when we went through not one, but two brutal storms. Storm Ciara turned the river into a raging torrent and, just as life was getting back to normal, we were hit by Storm Jorge, which brought blizzards. As if to prove that life goes on, no matter what is being thrown at us, right in the middle of this a new calf was born, who the children unsurprisingly decided to name Ciara.

In an ideal world the cows will have a natural delivery, and won't even need our help.

I'll drag myself out of bed at 2 a.m. to check on any expectant mothers, and again first thing when we get up, and sometimes in the night the cow will be standing there, happily munching on some hay, and when I am next back they have a calf up and about, and suckling.

That is the dream, but sometimes they need a hand and I'll pull on a rubber glove and feel inside, hoping that it is a pair of front hooves I touch, and that they aren't too big. Ideally the calf wants to be in a diving position, with their head facing forward and tucked in. Based on that information, and the mother's behaviour, we make a call on whether a natural birth seems possible, or if we need the vet for a potential C-section. It is stressful making this decision and important we know our own limitations, and get the vet if there is any doubt.

Once you make a call to start delivering naturally, there is no changing your mind and the worst thing that can happen is to get halfway and realize the calf is too big. We have a calving aid, a wonderful metal contraption with a ratcheting system that effectively gives you extra strength to help with the delivery of a bigger calf. The metal frame is placed on the labouring mother's hips and short ropes are attached to the unborn calf's front legs, allowing you to ease it through the birth canal. You need to exert just the right amount of force; the calving aid can be a dangerous piece of equipment in the

wrong hands, and can do more harm than good if used unsympathetically.

In the case of Ciara, her mother Margaret had no milk. She was elderly and had suffered from mastisis, but she was also our best cow, who gave us a calf every year, and we didn't want to part with her. Violet threw herself into the job of hand-rearing Ciara, who took to her and the other children, almost seeing them as her parents. It was lovely to watch her playing in the yard with them and chasing them on their bikes.

In 2021 Margaret once again produced a beautiful roan heifer calf, which we named Misty. We

As if to prove that life goes on, no matter what is being thrown at us, right in the middle of this a new calf was born.

knew there was no possibility that Margaret would be able to feed her, but it always pains me to take a newborn away from its mother, and I hoped we could leave Misty with Margaret but take charge of feeding her. Maybe this happy medium was achievable, I reasoned.

It wasn't; it really wasn't, as I soon discovered. Margaret had given birth to Misty with no problems, but was being less than maternal towards her newborn. She hadn't licked her, nuzzled her or interacted with her in any way. It is not unheard of for a new mother to be in a kind of shock after the trauma of the birth process, but Margaret started to become agitated when poor Misty began to take her first faltering steps, snorting, her tail swishing violently from side to side. Then, as Misty made her first attempt to suckle, Margaret swung around, bellowed, and knocked her calf to the floor.

Misty was mystified and, after clambering back to her feet, once again attempted to get close to her mother. This time Margaret got her head below her quaking, steaming calf and flung her into the air. She landed with a thud in the straw beside the wall, but before she had time to raise her head in surprise Margaret had come in for the kill and began to grind her into the floor, pushing her along

and into the wall. Misty was now making the most excruciating and pitiful noises.

'We've gotta do summat,' I was screaming.

'Yer can't go in,' shouted Clive. 'She'll 'ave yer.'

I set off running for my shepherd's crook, which was hanging up in the adjoining barn. By the time I got back, Clive was hollering at Margaret, who was taking no notice whatsoever, pretty much consumed with her mission to kill her calf. I climbed through the cow barrier adjoining the pen and reached through with my crook – I just wasn't close enough. Clive was now cursing me and Margaret.

'Yer stupid cow,' he yelled, and threw a pitchfork – the only weapon he had to hand. This only succeeded in making Margaret even more cross, and with a flick of her head she sent Misty flying in my direction.

This was it, my moment. Leaning as far through the barrier as I dared, I hooked my crook around Misty's crumpled torso and managed to roll her once over, close enough that I could grasp one of her back legs. It was ungainly, and certainly not the conventional way to handle a newborn, but it was the only option; I feared that there was a distinct possibility that Misty would be killed if left in situ.

Clive had by now climbed down to where I was hauling Misty up and through a metal barrier, and together we pulled her to safety. She lay, eyes wide and panting, while Margaret paced the pen looking angrily towards us.

'Ga' an get t'barrow,' I demanded. 'She's a pet now. Violet can look after her.'

'I just dunno whether she'll be alright,' Clive said. 'Could'a cracked 'er ribs and done all manner o' damage inside.'

We wheeled her out into the farmyard and told the children to stay well away from Margaret and to bed up a stable ready for their new pet, while I defrosted some colostrum. Fortunately Misty had sustained no long-term injuries, and she thrived. A full sister to Ciara, she too had a quiet, kind nature and was a joy to look after.

As for Margaret, after twenty-four hours of solitary confinement she calmed right down and reverted back to being the docile bovine that she usually is. I have a feeling that nature sometimes dictates so much more than we really understand; that somehow Margaret 'knew' that she could not feed her calf, and thus rejected her – though abandonment is one thing, infanticide another.

ABOVE The girls share out the milk so that there is enough to feed the bucket-reared calves.
OPPOSITE Princess is well equipped to withstand the cold Yorkshire months.

We used to have quad bikes, using them daily to go out on the feeding rounds, but they were both stolen. For a while we were without a bike, loath to get a new one for fear of it being stolen again. But I found I enjoyed going back to basics, riding the horses or going on foot. Using the horses to drive the sheep and cows was a new experience, both for freshly broken Princess and for me. She was certainly enthusiastic and would, if allowed, join the excitable cattle thundering down the field, bucking all the way.

Josie, our older mare, is intelligent, dependable and a natural shepherding pony. We have been all over the place as a team, moving cattle and sheep over roads and moorland. I realized just how much she had taken to it one day when we were driving some sheep back down towards a gate, and I had the reins in one hand, a crook in the other, and was focused on what Kate the sheepdog was doing. Some of the sheep began breaking away and, without any instruction from me, Josie worked out what was needed and began moving backwards and forwards in front of them to push them back into line. A bit like the sheepdogs, she wanted to be fully involved, and was enjoying her newfound role.

On occasion, some of the children will accompany me on the horses. Even Reuben came once, despite the fact that he is very clear that he is not a horse person. As it happened, there was method in his madness – he'd heard that it was a surefire way to impress the girls, although he was emphatic in his refusal to go the whole hog and don the jodhpurs.

Even Clive has warmed to the horses now they are brought back into work. I think he had previously seen them more as a frivolous hobby, responsible for eating everything and doing nothing! But once they became more useful as

a mode of transport, it was a game-changer. He is a big believer in a symbiotic relationship with the animals, where they need to be contributing to farm life to earn their keep.

The horses know the terrain well, and are so sure-footed that I always trust their judgement on which path to follow. If they deviate from the straight line in which I've headed them then there will be a reason concerning the state of the ground, and I let them take their lead. They were born on the moors and are familiar with the lie of the land, rarely making a mistake. Occasionally, though,

they will take a route that I wouldn't quite agree with, and seem to forget I am on their back . . .

We were gathering on the moor one day, with Clive and the dogs at the bottom of a ravine moving the sheep while I was at the top with Kate. It was a tremendous vantage point where we could stand guard, looking out over the heathered moortops. To be able to call this beautiful rugged terrain home still stirs a feeling of joy in me. As I sat admiring it, Josie's mane and my hair were blowing in the wind and I felt like the heroine on a movie set. Life was good. But Josie apparently wasn't feeling the same romanticism and suddenly decided she wanted to get involved in herding the sheep. Not only that, she was going to take the quickest route down. All I can say is, if you have ever seen the film *The Man from Snowy River*, where the main character, Jim Craig, goes down the most unbelievable incline, then you are just starting to get the picture. The only way I could stay on was by leaning right back, with my feet by her shoulders and my shoulders on her rump. I couldn't believe it when I reached the bottom and was still astride her.

'Did you see that?' I said breathlessly to Clive, full of exhilaration that I'd survived. What a horse-woman I must be!

'No, see'd what?'

We bring the yows down for an ultrasound scan near the end of the month, so we have a clear idea of how many are pregnant, and if they are expecting a single lamb, twins, or the occasional triplets. They are ready and waiting in the fields for the arrival of Adrian with his scanning machine. They are scanned one at a time, and marked on their back accordingly; a pop of colour, using sheep-marking fluid, that denotes their condition. Obviously we want as few geld yows – sheep who aren't carrying – as possible, and one lamb each is preferable to multiples. Single lambs can

be looked after by their mothers back at the moor, growing strong and healthy with little interference from ourselves, whereas twins require more sustenance from their mothers and must stay in the pastures.

I clearly remember during the 2020 scan it was snowing. A gentle snowfall of delicate flakes coming down around us, as we found out what our lambing session was going to look like. The year before, the children were nowhere to be seen, as it was so warm they were stripped bare and swimming in the beck. We don't let a balmy spell lull us into a false sense of security in February, but it's not easy to predict any more with weather patterns becoming increasingly erratic.

We are, as a family, acclimatized to our hill end, and don't really need it to be warm to swim. I have always been a fan of cold-water swimming – and just as well, as the water is pretty much always a varying degree of cold. There's a certain amount of satisfaction to be gained from braving the elements and wading out into the calm still waters of the tarn on a frosty day, when ice crystals cling to the pebbled shore and your extremities are soon numb.

I don't dive in, as that would be foolhardy and could cause shock. But I walk straight in, determined to keep going, let the pain of severe cold fade, then start swimming. Every bit of your body and mind is fighting it, as there is no denying how bitter it can be, but it feels rather like a personal battle of endurance, and overcoming the urge to retreat really gives you a mental boost. Once fully submerged, the cold is invigorating and heightens the senses, gifting you a moment of mindfulness while the tingling makes you feel alive. People say

ABOVE **Taff.**
OPPOSITE **Birkdale Tarn.**

you should do something every day that scares or challenges you, and this certainly ticks that box. It has become a more mainstream, fashionable pursuit now, rebranded as wild swimming. To me, it is just plain old swimming.

An active lifestyle makes for a healthy appetite, but with a large family to feed it can be hard to find the time to create a masterpiece at the dinner table. Fast food is what's required, but with our remote location it's not as simple as picking up the phone to the local take-away – not to mention how expensive it would be to order for eleven of us! Instead, we have over the years collected a few tried-and-tested recipes that are our own versions of all the favourites. A pizza dough is so easy to make, and everyone loves a pizza night, especially with sides of spicy potato wedges and a cooling raita dip.

I enjoy making an Indian meal too. It's not too difficult to make everything from scratch, from the naan bread – peshwari and its rich coconut flavours are the winner here – to chicken tikka and onion bhajis. I get rice in 25kg sacks, and have bags of chickpea flour, plenty of garlic and coriander, and a huge array of colourful and aromatic dried spices. On curry night the fragrant aromas of India waft across the farmyard and draw everyone in for dinner without me having to shout or ring the bell.

It is strange how fare once considered 'peasant food' can become popular. Curries, tagines, daubes and hotpots are all based on the idea of using poorer cuts of meat, then marinating and/or spicing, adding seasonal vegetables or grains and pulses to make a little go a long way. The effort comes in the initial preparation of the dishes, and then they can be left to cook gently for long periods, leaving plenty of time for busy people to get distracted by the thousand and one other jobs that need doing!

Onion Bhajis

If I say so myself, my bhajis are so much better than any I have ordered in a restaurant.

Prep time 15 minutes, plus resting time / Cooking time 20 minutes / Makes 8 bhajis

INGREDIENTS

300g onions

2 tsp salt

1 tsp ground turmeric

½ tsp ground coriander

½ tsp ground cumin

10g fresh coriander, finely
   chopped

1 small green chilli, deseeded
   and finely chopped

150g gram flour

1 tbsp vegetable oil, plus extra
   for deep-frying

Raita

½ cucumber, peeled and
   grated

150g Greek yogurt

10g mint leaves, finely
   chopped

salt and black pepper

METHOD

1. Halve then peel the onions and cut into 3mm slices. Pop into a bowl with the salt. Leave to stand for 30 minutes, then rinse off the salt and squeeze out any excess liquid.

2. Place the remaining dry spices, fresh coriander, fresh chilli and gram flour into a bowl, add the onion and mix well to coat the onion in the flour and spices. Add the oil and 4 tablespoons cold water and mix well until the ingredients come together.

3. Heat the vegetable oil in a fryer or deep-sided pan to 160°C. When the oil is hot, carefully drop 4 spoonfuls of bhaji mixture into the hot oil and cook for 4 minutes, until crisp and golden. Drain the bhajis on kitchen paper, then repeat with the remaining mixture.

4. For the raita, mix all the ingredients together in a small bowl and season to taste.

5. Serve the onion bhajis with the raita on the side.

Tip
If you like your bhajis a little spicier, then add the chilli seeds into the mixture.

Chickpea and Spinach Dhal

Prep time 10 minutes / Cooking time 25 minutes / Serves 4

INGREDIENTS

25g butter

2 tbsp vegetable oil

1 medium white onion, sliced
thinly

2 tsp yellow mustard seeds

½ tsp fenugreek

2 tsp ground cumin

1 tsp ground coriander

¼ tsp hot chilli powder

1 tsp ground turmeric

2cm piece ginger, peeled and
grated

3 cloves garlic, finely chopped

2 tbsp tomato puree

2 x 400g cans chickpeas,
drained and rinsed

200ml vegetable stock

400g carton of passata

150g baby spinach

1 small bunch fresh coriander,
finely chopped

1 tsp garam masala

METHOD

1. Melt the butter with 1 tablespoon of the oil in a large frying pan. Add the thinly sliced onion and cook over a moderate heat for about 5 minutes, until the onion is starting to soften and turn very slightly golden around the edges.

2. Add the mustard seeds, cover the pan with a lid, and wait a few seconds until you hear the seeds start to pop, then remove the lid and add the fenugreek, cumin, coriander, chilli powder, turmeric, ginger and garlic, and give everything a good stir for about 30 seconds.

3. Next, add the tomato puree and chickpeas to the pan, along with the vegetable stock and passata, then gently bring to the boil and simmer, uncovered, for about 20 minutes.

4. Add the spinach in batches until wilted, adding a splash of water if the sauce looks a little dry, then stir in half of the fresh coriander with the garam masala.

5. Serve the dhal in warm bowls with a sprinkling of the remaining coriander and some homemade peshwari naans (see page 54).

Tip
This recipe also works well with canned green lentils.

Peshwari Naans

One of Clive's favourite recipes.

Prep time 20 minutes, plus 1 hour to prove / Cooking time 12 minutes / Makes 4

INGREDIENTS

200g strong bread flour,
 plus extra for kneading and
 rolling
1 tsp fast-acting dried yeast
½ tsp sugar
½ tsp salt
1 tsp baking powder
100ml warm milk
50g natural yogurt
vegetable oil for greasing

Filling

40g flaked almonds
50g sultanas
40g desiccated coconut
2 tbsp soft brown sugar
50g soft butter

METHOD

1. Sift the flour into a bowl, add the yeast and stir to combine, then stir in the sugar, salt and baking powder.

2. Make a well in the centre of the flour, then pour in the warm milk and yogurt and stir with a spatula to combine into a dough. Turn the dough out onto a lightly floured surface and knead by hand for 10 minutes, until smooth. Return the dough to a lightly oiled clean bowl, cover with an oiled piece of cling film or a clean damp tea towel, and leave in a warm place for an hour until the dough has doubled in size.

3. Mix the filling ingredients together in a small bowl, then divide the dough into four balls. Flatten out each ball, then divide the filling and place into the middle of each disc. Pull up the dough to seal the filling inside, then carefully roll on a lightly floured surface into a tear shape about 20cm long by 12cm at its widest part.

4. Heat a large non-stick frying pan and dry fry the naan bread over a moderate heat for about 3 minutes per side. Repeat with the remainder of the naans and keep warm.

Tip
For variety add some garlic and chopped coriander to the dough.

March

The first sign that spring is on its way, and that winter has finally released us from its icy grasp, is the appearance of a little sprout of a plant that we call moss, but is really the emerging dark green leaves of cotton grass. Unlike the migratory birds, who can be deceived by warm weather into coming back too early, this succulent arrival is a reliable harbinger of good times, and that is why we love it. The countryside is awakening, and there should now at least be the chance of peeling back a few layers and consigning the heaviest of our winter wear to the hall robe cupboard.

It turns out that the sheep love the appearance of cotton grass even more than we do. They find the young sweet plant absolutely irresistible. Any place where the moss pops up, generally in the tussocks that are to be found beside marshy ground and on rockier outcrops, will be hotly contested by sheep from all around the area. It is the one temptation that can take them away from their heaf, their ingrained respect for those invisible boundaries disappearing when a mouthful of moss is the prize.

Left alone, the tiny plant would develop a fluffy white seed head over the summer, which over the years has been used for everything from stuffing pillowcases to making a medicine to treat diarrhoea. But to reach the stage of those distinctive heads, it has to survive many a hungry and adventurous sheep. We have to keep an eye and regularly gather them back to their own area as, with April and lambing time approaching, it isn't a good time to lose track of any heavily expectant yows. Inevitably the lure of the moss occasionally wins, though, and the odd mother will reappear in late April, a lamb in tow. Sometimes the wanderer will appear without a lamb, and you have to assume they have succumbed to one of the moors' many bottomless bogs that are amongst the worst hazards for our sheep. It is one of the reasons why, when people ask how many sheep we have, I only ever feel we can give an approximation, as it can be slightly fluid. Over such a large expanse, you cannot possibly know the whereabouts of all your flock and, although their welfare is paramount, there is a certain amout of risk that a hill sheep will inevitably encounter. You cannot take away all danger.

The month is all about keeping an eye on the yows and making sure everything is ticking along smoothly for them, while preparing for lambing time, with tasks such as getting the lambing barn and hospital ready.

Although I consider myself extremely fortunate to live at Ravenseat, I am always aware of the fact that we don't own it. The farm is part of a 36,000-acre estate owned by an American billionaire entrepreneur called Robert Miller. We have a tenancy agreement, but not the security that we would entirely wish for, and this is one of the reasons we bought The Firs: as a place the family could put down lasting roots. It, too, was a farmhouse in its own right, but was sold by the previous estate owner. Most of the land that went with The Firs was sold too, so now it can only be classed as a smallholding with a few acres surrounding it. But, nevertheless, it belongs to us.

One day in March 2020 I was driving home when my attention was caught by a man hammering in a For Sale sign by the side of the road. Further investigation revealed that Smithy Holme Farm was going up for auction. This was a turn-up for the books.

Smithy Holme used to belong to a friend of ours, Tot, who for a while had lived in a caravan on the land, the old crumbling farmhouse too much for him to face repairing on his own. He tended to his sheep every day, even when he had moved into sheltered accommodation. It felt like the end of an era when he died, a link with the past and a generation that worked the land through the war years and beyond. He saw such great changes in society, and a shift in farming and country life, but remained positive and upbeat right until the very end.

According to the listing, the auction lot included the farmhouse, known as High Smithy Holme, or Anty Johns, with planning permission already in place, thirty acres of land, barns and the right to graze 180 sheep on the moor. We had looked at Smithy Holme when it came up for sale a few years earlier, but felt at the time that we

lacked both the skills to take on the renovation work ourselves, and the time to handle such a big project. It had been bought by a local farmer, but was now back on the market and this time, having successfully renovated The Firs by then (admittedly a much smaller project), we decided that we would try to buy it. It would be an amazing opportunity to own a farm of our own, with grazing rights. Coupled with The Firs, it felt like we were turning back the clock; instead of seeing the gradual fragmentation of these little hill farms, we were doing the opposite and putting one back together. Reclaiming history, I suppose.

Anty Johns may be in need of a lot of love, but it has a character all of its own that we felt was worth saving.

I knew the farmhouse was seriously dilapidated. The roof had partially caved in due to rotting timbers, and although the exterior walls stood good and strong, the sight inside of a rubble-strewn floor and crumbling plaster brought home the full extent of the work that would be involved in making the place habitable. There was no water or electricity, and the track to it was steep, rutted and in parts had a precipitous drop alongside – but I felt that this inaccessibility was part of its charm. The beautiful cobbled yard, old fireplaces, fell rights and history of the place all spurred us on to see past the overwhelming difficulties, and gave us a determination to resurrect Anty Johns and bring it back to its former glory.

Having land that was completely unfamiliar to us meant new places to explore and understand.

The Covid-19 pandemic and the fact the entire country was in lockdown led to the auction being conducted via the telephone. I sat at the kitchen table with the phone pressed to my ear, nervously listening to the auctioneer's usual patter, feeling both exhilarated and sick in equal measure. I knew both the guide price and my limitations, and as the bidding rose in increasingly large increments, I became more and more unsure I was making the right decision. The bidding slowed; I held off and listened as the auctioneer offered it up once again, going once, going twice . . . it was now or never.

'I'm in,' I said, and with that the gavel fell.

I had bought it!

Having land that was completely unfamiliar to us meant new places to explore and understand. We know the acres that surround us at Ravenseat, our home. The farmhouse in which we reside is the hub, the nerve centre, but it is the rolling fields and bracken-thick slopes surrounding us that really resonate with the soul. This connection we feel to the land is not a product of ancestral inheritance – that Clive and I both come from families outside the farming community is testament to that. It is an openness and desire to feel a sense of belonging that really matters, a need to know and understand an age-old system and to recognize the values that traditional farming can hold for us in modern-day life.

So now it was time for us to learn about Anty Johns. We had maps and paperwork detailing field areas and land codes, but nothing that really told us about the human element. What were the field names; could we find the natural springs; where did the sweetest grass grow; which way did the wind blow? We had been to view what was on offer prior to sale, but now we could have a proper poke around without feeling we were trespassing. Anty Johns is our version of next door, a mile-and-a-quarter walk through the Close Hills Pastures following what is now the Coast to Coast footpath, but was originally the main route from Keld to Ravenseat before the road was put in.

A procession of children, terriers and the 4x4 utility vehicle, fully laden with a picnic, made its way towards the new house. The first issue we encountered was the boundary moor gate between the pasture and the moor. Outwardly it looked wide enough to get the 4x4 through, but it soon became apparent that it was narrower than we had realized. Also, I had become acutely aware that just inches from the gate was a precipitous vertical drop down into the Boggle Hole.

The children who were riding in the back disembarked – a sure sign that this was going to be a precarious move. Clive was now standing in front of me on the other side of the gate, huffily swinging his arm in a windmill motion.

'Come on,' he shouted impatiently as I edged forward. 'Yer could get a bus through there!'

I felt less convinced.

'Back up, yer not lined up properly,' he shouted.

Flustered, I put the vehicle into reverse, but my mind was more focused on how many seconds it would take until I hit the bottom of the ravine.

With a jolt, my back wheel went over the edge. I froze, hardly daring to breathe, convinced my weight was now the only thing preventing the 4x4 and me from plummeting to an untimely death. As I tried to work out what to do, it became apparent there is only one thing worse than hanging with one wheel in mid-air over a precipice, and that is hanging with one wheel in mid-air over a precipice with an irate husband shouting instructions!

'Get out,' huffed Clive, who appeared more angry than concerned. I tentatively undid my seatbelt and made my escape. I have to say that, once back on terra firma, the position of the car didn't look anywhere near as terrifying as it had felt – but I was happy to be handing over control to Clive.

At least we had worked out our first job – widening the gate.

As quaint as it was to risk life and limb driving to Anty Johns, access was clearly more of an issue than I'd anticipated – as I discovered when my plans to get into the fields and cut them back for some beautiful hay bales later in the year were also scuppered. A narrow packhorse bridge is the only way to access the track and, while thoroughly charming, isn't conventional baler-friendly. Widening a bridge was certainly not an option, so those plans were put on hold. A more compact round baler would have to be used until we found a smaller conventional version. For now, though, I decided to focus on finding a water supply to feed the house.

There is a water collection point that would have sufficed for past residents (at least those who had never luxuriated in indoor plumbing), but

It may have a big hole in the roof, but you can see how solidly built the farmhouse at Anty Johns is.

what I wanted was to find a freshwater spring. Ravenseat, like all the properties at the top of the dale, is supplied by spring water that travels from a tank a mile away on the moor, via another tank and into the farm, with a charcoal filter to clean it. The water pressure is great, and it has only failed us once, when the spring inexplicably moved.

Although there was always the option of drilling a borehole, we decided to try to locate a spring ourselves with copper divining rods. I wasn't entirely convinced they would be effective, but the children were enthusiastic, so we conjured up all our mystical energies and set off in search of our little oasis. I have to say that the divining rods seemed to work – though perhaps almost too well,

ABOVE **An old lime kiln at Anty Johns.**

as they were constantly crossing downwards to indicate what seemed to be water sources under the surface everywhere. I wasn't sure how exactly to specify that I needed a bubbling, clear, cool freshwater spring, rather than muddy saturated bog puddles.

We needed something more tangible, and eventually found water in pure rock that flowed into a small cavern. Hidden at one end of the limestone escarpment beside the beck was an unassuming fissure in the rock . . . the entrance to Brian's Cave.

I had seen references in books to this most underwhelmingly christened place. Interestingly, in both instances the writers were unable to actually locate Brian's Cave. One had almost certainly been given misinformation as to its whereabouts, the intrepid nineteenth-century explorer having been told that it was to be found up in Whitsundale, which is a couple of miles away.

From the outside we could hear the faint echoes of water falling. I leaned forwards, stretching as far as I could and skewing my body sideways so I could fit through the narrow entrance.

'Git yerself reet in an' 'ave a look?' suggested Clive.

I shuffled a little more, brought my legs around beneath me and shimmied sideways between the sharp protruding rocks, and finally slipped through the gap. I was thankful to be in trainers as opposed to wellies for once, needing the extra dexterity. Once through and around the corner, the cave opened out a little. I edged along cautiously, using my phone as a torch. Looking upwards into semi-darkness, I soon realized that it

was probably not such a great idea to be potholing without a helmet and proper equipment; the roof above me looked unsafe, to put it mildly, with jutted, angular rocks, and numerous small streams of water cascading down into a shallow pool under a precarious overhang. Until then I had been filled with complete wonderment at this subterranean masterpiece, but now I was overcome with claustrophobia. I needed to get out, quickly and quietly, my mind filled with thoughts of noise vibrations sending the rocks overhead thundering down.

'Are yer alreet?' hollered an all-too-familiar voice, reverberating through the cave. 'I said ARE YER ALREET?'

I felt quite sick, and hastily made a backwards retreat.

Breathing fresh air felt jolly good after the confines of Brian's Cave, but it did seem that at least my efforts were not in vain and we'd achieved what we'd set out to find: an underground water source.

Although for modern-day living we need Anty Johns to have the amenities, the aim still remains to recreate and reassemble the original features of the old farmhouse, keeping it true to its Yorkshire heritage. A cartographer called Anthony Clarkson lived there in 1800. As a learned gentleman (he'd been a schoolmaster previously), he was tasked with drawing up the tithe enclosure maps in the area and, as a keen diarist, he also wrote extensively about daily life in Swaledale. His notes, where he talks of his life and his travels around the area, sometimes resemble an epic pub crawl through places that we could drive to in an hour but which would take him a day to reach on horseback, or longer on foot. It made me smile to see Clarkson complaining about the state of the farmhouse, and needing to do some work on the leaking roof. Here

I am 200 years later thinking exactly the same thing – it goes to prove that life here really doesn't change.

We don't yet have a long-term plan for who will live at Anty Johns. As long as we have Ravenseat we will stay put, and perhaps it will be home for one of the children one day. They are all working out their own routes into the world, and I'd never push any of them into a particular career, but Sidney and Miles are both showing a keen interest in being farmers. So, who knows, maybe one day they will be able to live there.

As the weather begins to warm a little, so we find ourselves inundated with eggs from the chickens. We have around fifty hens and a few cockerels, who wander freely around the farm, occasionally making a nuisance of themselves by roosting in unsuitable places such as above the cab-less David Brown tractor, leaving hen manure on the seat. The majority of the hens are ex-commercial layers that we collect from a local farmer who offers them up for rehoming when egg production slows down. They are still tremendously fruitful and keep us in eggs for the majority of the year, the surplus of which we can sell at the farm gate.

Miles is in charge of the chickens and collecting the eggs, and takes great care of them. He knows everything about each individual chicken, from its age to its family tree. Point at one, and he'll tell you all about it.

There is no such thing as too many eggs and they make regular scrambled appearances at breakfast, or in eggy bread or French toast. It's incredible how many you can get through on baking day, and 'We need to use up some eggs' is an oft-quoted excuse to make yet more cakes, chocolate brownies or other sweet treat.

Eggs keep fresh for much longer than people realize, and obviously have a longer shelf-life when you have them straight from the hen. Cut out the grading and packaging process, and they might still be warm when you crack them into water for the perfect poached egg. No need for any vinegar in the simmering water – a sign of freshness is that the yolk and white hold together without it.

We have peafowl too, though more for ornamental than breeding purposes, as they are such terrible parents it's a miracle that the entire breed has not died out. The biggest issue seems to be that they get bored of sitting on their eggs, and stubbornly refuse to last the duration. A better option by far is to put any peafowl eggs under a broody hen. Not only are they more likely to hatch the eggs but they make much better mothers all round, having far stronger maternal instincts!

Alternatively, the eggs (as sizeable as a goose's) are rather good baked in the oven with a dash of cream and a grinding of pepper.

Leek, Spinach and Goats' Cheese Tart

By March the hens have started laying after their winter break and we have a glut of eggs. This tart is a great way to use some of them up.

Prep time 20 minutes, plus 10 minutes chilling / Cooking time 55 minutes / Serves 4

INGREDIENTS

Shortcrust pastry

175g plain flour, plus extra for
    rolling

50g butter, diced, plus extra
    for greasing

50g lard, diced

1 egg yolk

Filling

20g butter

250g leeks, trimmed and
    sliced

150g baby spinach leaves

3 medium eggs

100ml double cream

125g soft goats' cheese, sliced

100g roasted red pepper,
    diced

salt and black pepper

METHOD

1. For the pastry, sift the flour into a bowl, add the butter and lard and rub together with your fingertips until the mixture resembles fine breadcrumbs. Stir in the egg yolk with 3 tablespoons cold water. Bring the pastry together with your hands, then wrap in cling film and chill in the fridge for 10 minutes.

2. Put a baking tray on the bottom shelf of the oven and preheat the oven to 200°C/180°C fan/gas 6. Grease a 23 x 2.5cm tart tin with butter. Roll out the pastry on a lightly floured surface and line the tin, then prick the base and sides of the pastry with a fork. Leave a little pastry overhanging to allow for any shrinkage when it is cooked.

3. Line the pastry case with baking parchment and fill with baking beans. Cook for 15 minutes, then take the beans out and cook for a further 5 minutes. Allow to cool slightly, then trim off any excess pastry.

4. Meanwhile, to prepare the filling, melt the butter in a frying pan, add the sliced leeks and cook gently for 8 minutes until softened. Add the spinach to the pan in handfuls and stir gently for 3–4 minutes until the spinach wilts. Season with salt and black pepper and set aside to cool.

5. In a jug, whisk the eggs and cream and season with salt and black pepper. Place the leeks and spinach in the bottom of the pastry case, top with the goats' cheese and red pepper, then pour the egg mixture carefully over the filling.

6. Lower the oven temperature to 190°C/170°C fan/gas 5 and carefully transfer the filled tart onto the hot baking tray. Bake the tart for 35 minutes, or until golden and softly set.

Tip
As an alternative, try adding feta cheese. Chargrill your peppers or, for convenience, use ready-prepared roasted peppers.

Yorkshire Curd Tart

This is one of my all-time-favourite desserts. Cottage cheese is the easiest cheese to make, if you fancy going one step further and preparing it from scratch rather than buying it ready made.

Prep time 20 minutes, plus 10 minutes chilling / Cooking time 1 hour / Serves 4

INGREDIENTS

Sweet shortcrust pastry

115g plain flour, plus extra
   for rolling

85g butter, plus extra for
   greasing

20g icing sugar

1 egg yolk

Filling

115g butter

115g caster sugar

2 medium eggs, plus 1 egg
   white, beaten together

200g cottage cheese

50g currants

pinch of ground nutmeg

zest of 1 small lemon

METHOD

1. For the pastry, sift the flour in a bowl, add the butter and rub together with your fingertips until the mixture resembles fine breadcrumbs. Stir in the icing sugar, then the egg yolk. Bring the pastry together with your hands, then wrap in cling film and chill in the fridge for 10 minutes.

2. Put a baking tray on the middle shelf of the oven and preheat the oven to 200°C/180°C fan/gas 6. Grease a 21 x 2.5cm tart tin with butter. Roll out the pastry on a lightly floured surface and line the tin, then prick the base and sides of the pastry with a fork. Leave a little pastry overhanging to allow for any shrinkage when it is cooked.

3. Line the pastry case with baking parchment and fill with baking beans. Cook for 15 minutes, then take the beans out and cook for a further 5 minutes. Allow to cool slightly, then trim off any excess pastry.

4. To prepare the filling, beat the butter and sugar until soft, then add the eggs a little bit at a time, until combined. Add the cottage cheese and lightly whisk to break up any large lumps. Once blended, add the currants, nutmeg and lemon zest. Pour the curd mixture into the pastry case and cook on the middle shelf of the oven for 40–45 minutes, until the filling is set, but still has a wobble.

5. Allow to cool in the tin before slicing.

Tip
The pastry can be prepared in a food processor. Add the flour, butter and icing sugar, pulse until the mixture resembles breadcrumbs, then add the egg yolk and pulse until the pastry forms. For a delicious twist try adding a few drops of rum to the filling.

April

'By hook or by crook' is an oft-used term, the origins of which date back to the Middle Ages. It supposedly refers to the right of the people to take dead wood from common land for their fires – there was to be no cutting of trees, just the salvaging of any dead wood that could be reached with either a shepherd's crook or an agricultural labourer's weeding hook.

It is during April that my shepherd's crook comes into its own. I know that it is seen very much as a prop that symbolizes the role of a shepherd, but it is in fact an essential tool – and never more so than during lambing time. Combine a crook with an able and willing sheepdog and you are a force to be reckoned with when it comes to catching yows and lambs.

Most of our sheep will lamb outside in the fields, having been brought down from their heafs at the moor, as we prefer them to give birth outdoors, in a natural environment where there's less risk of infection and mis-mothering. Invariably some will have to come inside into the buildings, particularly shearlings (first-time mothers) expecting twins, as they are more likely to be in need of assistance. Once inside they need to be carefully watched over throughout the day and night, for as soon as a yow has lambed she and her offspring need to be moved into an individual pen to avoid any confusion with newborn lambs bonding with the wrong mothers and then being rejected.

Favourable weather conditions make for a successful lambing time, though opinions vary as to what is climatically perfect. I would maintain that spring 2020 was as near perfect as can be – it was dry but not too warm, and there was still a nip in the air at the break of day. Weather that is too hot can lead to ailments such as rattle belly, also known as watery mouth. Lambs afflicted have a drooling mouth and take on a strangely pot-bellied appearance. On the other hand, if it rains a lot and the land is saturated then the lambs are at risk of joint ill, a dreadfully painful condition caused when bacteria manages to get in through their wet navels.

A windy, cold snap is not so harmful to the lambs but can be the cause of mastitis in the yows. The udder becomes sore, and consequently the yow stops her lamb from suckling, leading to more inflammation which can result in necrotizing tissue, infection and a poisoned system, which can be fatal. It is known in these parts as 'black bag' and can be recognized early on by the awkward gait of a yow that is clearly uncomfortable. Thankfully these troubles are few and far between, and lambing outdoors is certainly a healthier

In a small, uniform field with a stream running through it there would be no real problem locating the lamb, but it becomes more of a challenge in a fifty-acre undulating pasture.

environment – though you are at the mercy of Yorkshire's unseasonable and changeable weather.

Fortunately we didn't have to contend with many of those ailments during the 2020 lambing season, and in fact the main issues we faced arose when the ground began to dry up – not something we ever expect to deal with on our wet, boggy moorland.

A lactating animal needs plenty of access to liquid. A cow who is being milked daily will drink gallons of water, and the sheep are the same, so it is very important to keep them hydrated. For the first time we could recall, we were having to put buckets of water out into the fields. There is not a single field that has a water trough; all have natural springs or streams that provide fresh, clean drinking water for the stock. When these dried up and there was no rainfall for weeks on end, our only option was to walk down to Whitsundale beck, a river that never runs dry, and form a human chain of children carrying buckets to pass over the walls or fences. As quickly as we filled buckets, the yows drank them.

Hydration is vastly important not just for the health and well-being of the yows, but also their lambs. A dehydrated yow will produce incredibly rich milk, which has an unfortunate and immediate effect on the digestive system of a newborn lamb – constipation. A 'pinned' lamb, for this is what we call it, cannot thrive when their tail has become impossibly welded to their backside with sticky bright yellow poop, which in turn prevents any further movement of the bowels. It must be an uncomfortable problem, and one that will prove fatal if not spotted quickly. It is easily cured by gently pulling the tail upwards. The relief is instantaneous and the result is often spectacular.

Thirsty yows in the allotments and high pastures have, in the past, resulted in us encountering a few frustrating mysteries. We would spy a lone yow displaying all the telltale signs of having lambed, maybe a little blood around the tail and with a hollow look about her torso. Sometimes just her mannerisms would alert us, depending on her maternal qualities. But where was her lamb?

The very first thing that a yow wants to do after her labour and birth is to have a drink. She will set off with her newborn lamb, unsteady on its legs and blindly following its mother, down to the water's edge. Ever creatures of habit, the sheep know where to access their drinking water, but with water levels low, they sometimes have to go right down the steep banks and into the gullies to be able to get a sup. The poor lambs do not yet have the strength to climb back up the steep sides. The better, more attentive mothers can be found standing guard over their lambs, looking down as though marking the spot, and in these cases it is straightforward enough to go and fish her offspring out and reunite the fond pair. Other mothers aren't quite so maternal and, unperturbed

at the loss of their nearest and dearest, will head for the hills alone.

In a small, uniform field with a stream running through it there would be no real problem locating the lamb, but it becomes more of a challenge in a fifty-acre undulating pasture with cliffs, valleys and ruts, criss-crossed by rush-covered drainage systems. As a consequence there are mysterious disappearances that we will never solve. Just recently a cow came down to the farm and had clearly calved, but there was no calf to be found. We searched everywhere, even getting the dogs to see if they could track it down or pick up a scent. The children used fishing nets to check the deeper pools of water, but still our efforts proved fruitless. Maybe we will find its remains at some point. Perhaps it was born dead, or drowned and was washed away.

One morning Clive came back in from his early rounds with a yow that had been scanned for twins. She had clearly lambed, but her new offspring could not be found anywhere. After searching for a good while he accepted defeat and brought her down to the farm and gave her a pet lamb – an orphaned or multiple who we bottle-feed until we can find a new mother. There is little point in wasting a yow's milk and it provides a better situation for the pet lamb than hand-rearing. This yow was somewhat ambivalent about the adoptee, ignoring the bleating lamb with which she now shared a small pen. She seemed very much aware that it had nothing to do with her. In this situation

From left to right: Kate,
Nell, Midge, Bill and Taff.
The younger dogs learn
so much from the older,
experienced ones.

the only way that you can try to convince her to accept the lamb as her own is to rub it under her tail and hold her still while it suckles. It can be a very long, drawn-out process, but if you are determined then you will usually win the yow over . . . eventually.

Generally Clive and I look after our own heafs, as familiarity with the land and the flock that you tend is of great benefit. We know the places the lambing yows favour and where to look for trouble, but that particular day had been less frantic than usual so I went with him to help out. At the top of the moor, and with Kate at my side, I stopped for a moment to look in a southerly direction to the pastures where I could see my

flock peacefully grazing. All was quiet, other than a distant melodious warble of skylarks carried on the gentle breeze. It was utter tranquillity. Then I heard the feeblest of muffled bleats; I glanced at Kate, whose ears were pricked, her nose in the air. Looking aloof, she was stock still, listening hard. After another minute or so we both heard the noise again, and now I saw recognition in Kate's eyes as she looked up at me expectantly to see what was required. To have a good sheepdog is to have a real understanding between you, a relationship whereby the sheepdog doesn't just respond to each individual command, but comprehends the bigger task and what you are trying to achieve. Ideally when moving the flock you don't want

When finally I reached Kate, sure enough, there were two muddy, pitiful little bundles standing quaking in a very well-hidden underground drain.

to be repeatedly having to tell the sheepdog to 'come by' and 'away'; you want a dog that is clever enough to know what is happening and does all the necessary work.

There is, obviously, no command for 'can you locate the faintly bleating lamb?', but Kate knew exactly what was needed. I nodded permission and off she went down the slope, nose down, looking into the gully from time to time. I stayed where I was, having the perfect vantage point to see Kate working, trying to pick up a scent. About a quarter of a mile on, she suddenly stopped. Her shoulders and ears forward, she circled around, her tail erect; she was on her toes knowing that she had found the prize. I scooted down towards her,

half-scrambling and half-sliding on my behind. When finally I reached Kate, sure enough, there were two muddy, pitiful little bundles standing quaking in a very well-hidden underground drain. I got to my knees, finding that I could only just reach down far enough to pull them out. They were cold and hungry, but otherwise alright. They had been extremely lucky – any more water in the drain and they would have either drowned or succumbed to hypothermia.

I had to break the news to Clive that he was going to have to un-mother his pet lamb, which he duly did. The biggest surprise was that the yow that had absconded in the first place, abandoning her newborn lambs in the underground drain, took them straight back without issue.

All the children enjoy helping out at lambing time. Annas checks on newborn twins that she helped to deliver (left) and Clemmy feeds a pair of twins warming up on the hearth (above).

I would never say that a sheepdog is a vital tool of the trade, as truly this does not bestow enough of a value upon what is essentially a partner, friend and colleague. The relationship that is built between the shepherd and sheepdog ultimately dictates how you work together as a team. Each dog has a unique personality, and thus will have strengths and weaknesses in the same way that a person might excel in one area but not in another. To get the most out of the dog this must be recognized and their strengths nurtured. Some dogs will work for anyone, others only for their shepherd; some respond to voice command alone, others only to a whistle, some both.

At times we operate at a great distance, particularly when we gather the sheep in from Whitsundale. You can have a jolly decent view of where the sheep are if you are standing on the ridge at the opposite side of the valley, though the actual task of moving them towards home then falls to the sheepdog. At first, sheep and dogs are just specks on the far slopes and screes. The sheep are

easily able to escape from the clutches of a sheep-dog that has already had an outrun of anything up to half a mile to reach them, so it is now that the wisdom and intelligence of the sheepdog is relied upon. They must have an understanding of what the goal is, where we want the sheep and how to go about this task, as we are too far away to be of any physical help. We want the dogs to move the sheep along at all costs, not to baulk if one of the yows refuses to budge. Sometimes a nip at the heels is the most effective encouragement. This breach of the shepherds' code would result in disqualification at a sheepdog trial – but out in the field, at the moor, this is exactly what is needed.

Working with livestock can give you a fresh perspective on key moments in your own life.

We think of ourselves as intrinsically different to animals, but tending our livestock on a daily basis, year in, year out, it does become abundantly clear that, when it comes to the basics of life, there are so many similarities. Giving birth is a fine example. Attending labouring animals and assisting when required definitely demystifies the process and provides insight into the body's natural abilities. It also highlights the issues that might arise and how to rectify problems – and, crucially, demonstrates that you cannot plan for anything. A calm and pragmatic overview will serve you well, and is the best way to go through life.

OPPOSITE 'Nothing to see here.' Ciara's passion was searching for socks and underwear to chew upon.

Creamy Ham and Cabbage Pie

I love this pie as a hearty dinner but also eaten cold the next day. Everyone tends to sort out their own lunches, and can cut off a slice as they want, for a filling midday meal. People talk about boiled hams, but the key for me here is just letting it gently simmer.

Prep time 30 minutes / Cooking time 1 hour 50 minutes / Serves 4

INGREDIENTS

1kg gammon joint

1 onion, roughly chopped

2 bay leaves

½ tsp black peppercorns

Pastry

170g plain flour, plus extra
    for rolling

85g butter, cut into cubes

¼ tsp English mustard powder

40g strong Cheddar cheese,
    finely grated

1 egg, beaten

Sauce

30g butter

1 small leek, sliced

2 cloves garlic, crushed

200g sweetheart cabbage,
    shredded

25g plain flour

200ml hot chicken stock

200ml crème fraiche

1 tbsp Dijon mustard

salt and black pepper

METHOD

1. Sit the gammon in a large pan with the onion, bay leaves and black peppercorns, then cover with water. Bring to the boil and simmer for 45 minutes.

2. Remove the gammon from the pan, allow to cool slightly, then cut away the skin, leaving a thin layer of fat attached.

3. Preheat the oven to 180°C/160°C fan/gas 4. Bake the gammon for 30 minutes in a small roasting tin. Allow to cool, discard the fat and shred the meat.

4. Make the pastry while the gammon cooks. Sift the flour into a bowl and rub in the butter with your fingertips until the mixture resembles fine breadcrumbs. Stir in the mustard powder and cheese, then add 2–3 tablespoons of cold water and mix until the pastry comes together. Wrap in cling film and chill in the fridge.

5. For the sauce, melt the butter in a frying pan. Next add the leek, cook for 5 minutes until softened but not coloured, then add the garlic and cabbage. Continue to cook for 2–3 minutes until the cabbage has wilted.

6. Sprinkle the flour over the vegetables and cook out the flour for 2 minutes, then pour in the stock, crème fraiche and the mustard. Gently fold the ham into the sauce and season to taste with salt and pepper. Simmer for 5 minutes until the sauce thickens.

7. Transfer the filling into a 25cm deep pie dish and preheat the oven to 220°C/200°C fan/gas 7.

8. Roll out the pastry on a lightly floured surface until it is large enough to cover the top of the pie dish. Dampen the edges of the pie dish with water, then lay the pastry over the filling, trim off any excess then crimp the edges. Make a small insert in the centre of the pie for the steam to escape, then brush the pastry with beaten egg.

9. Bake for 30–35 minutes, until the pastry is golden and the filling is piping hot.

Rhubarb and Custard Crumble Cake

On a good year the rhubarb will have appeared and be ready to harvest, otherwise I have to hope I still have some in the freezer. But you can never go wrong with a crumble and when you make that into a cake, it's even better!

Prep time 15 minutes / Cooking time 40 minutes / Serves 4–6

INGREDIENTS

Rhubarb

250g Yorkshire rhubarb, cut
    into 2cm chunks
50g caster sugar

Crumble

50g plain flour
25g butter, cut into cubes
40g chopped mixed nuts
25g demerara sugar

Cake

170g butter, softened, plus
    extra for greasing
85g caster sugar
85g soft brown sugar
3 medium eggs, beaten
1 tsp vanilla extract
170g self-raising flour
1 tsp baking powder
30g custard powder
icing sugar to dust
custard to serve

METHOD

1. Preheat the oven to 180°C/160°C fan/gas 4. Grease and line a 21cm round springform cake tin.

2. Put the rhubarb into a saucepan and sprinkle over the caster sugar. Heat very gently until the sugar dissolves into a syrup. Simmer for 6–8 minutes until the rhubarb just begins to soften. Set aside to cool.

3. Meanwhile make the crumble. Place the flour into a mixing bowl and rub in the butter with your fingertips until it resembles breadcrumbs, then stir in the chopped nuts and sugar.

4. Next make the cake. Beat the butter, caster sugar and soft brown sugar in a bowl using an electric mixer for 2–3 minutes, until light and fluffy. Add the eggs and vanilla extract with 2 tablespoons of the flour, beat for a minute until combined, then fold in the remaining flour, baking powder and custard powder.

5. Drain any excess juices from the rhubarb and discard the syrup, then gently fold the cooled rhubarb through the cake mixture. Transfer the cake batter to the prepared tin and gently smooth the top with a spatula. Sprinkle the crumble over the cake and press down lightly with the back of a spoon, then bake on the middle shelf of the oven for 40 minutes, until the cake is golden, firm to touch, and when a skewer is inserted it comes out clean. Allow to cool slightly in the tin, then transfer to a plate.

6. Dust lightly with icing sugar before serving warm with custard.

Tip
The cake is also delicious served with ice cream or single cream.

May

'Ne'er cast a clout 'til May be out.' Nobody really knows whether the May referred to in the traditional saying refers to the May hawthorn blossom flowering or the ending of the month but as both seem to coincide in the Upper Dale there's little to mull over. Inevitably layers of clothing will be put aside as the month nears its end, although there will be an occasional sharp morning frost that calls for jackets to be rooted out once more. But once the morning's haze has lifted there will be days of warmth and a gradual greening of our surroundings – the new fronds of bracken unfurling, the trees now all in leaf.

Traditionally, 1 May is grass day, the day that the cattle are released from the confines of the cowsheds and moved into the allotments. Finally, after months of us bedding up their byres and carting hay and silage up the fodder gang, we throw open the doors for the last time. Unaware that their moment of freedom has arrived, the cows dutifully amble into the yard, as they do on a daily basis while we muck them out. They mill around the gate in their usual indifferent manner, occasionally scratching their necks on the corner of the stone barn, undoubtedly itchy with an accumulation of dry scurf, hay seeds and dust. Encrusted muck buttons cling to their underbellies; like us

they'll cast off their coats when the weather allows it, replacing the thick fluffy winter covering with a finer, glossy, summer sheen.

When it is time to open the gate, indifference is replaced with a palpable excitement, and every cow suddenly wants to go through first, eagerly jostling for prime position. We have to man the gate and let them through carefully, in as orderly a fashion as possible, or risk the wall cheeks coming down. As the evening approaches, the novelty of being free outside wears off. Routine means the cows will expect us to visit with some food, so we put hay out for them to help them make the connection that the outdoors is now home until the autumn.

Only the house cow, Buttercup, will have a different routine, tempted to the gate each morning and evening with the rattle of a bucket of food, so that we can milk her. She gives her best milk throughout the summer months, thanks to the diet of sunshine and grass.

By the middle of June the meadows are cleared of sheep, the lambs all tagged and pedigrees recorded. Strokes and pops of coloured marking are carefully applied to the fleeces of the lambs, distinguishing ownership and the heafs on which they belong, and then they, along with their

mothers, move either to the allotments and high pastures (for yows with twins) or back to the moor if a single.

The meadows are now 'closed up' and entirely given back to nature for the next couple of months. It is a pleasing task indeed to quietly walk the sheep back to their respective heafs for summer grazing. Through the meadows we must go, the dogs weaving back and forth, nipping at the heels of the sheep that lag behind, chivvying them along through the grass. The movement of the flock creates a disturbance underfoot and a cloud of insects rise to the sky. Looking up we see swallows swooping and diving as they seize on the opportunity to feast on the tiny invertebrates.

'Turning away' is the moment when the fruits of our lambing time labours become apparent, the culmination of weeks of toil, or months if you include the seemingly endless winter when we tended the flocks daily whatever the weather. It's a turning point indeed, where both we and the sheep can unleash ourselves from the shackles of the daily grind. They no longer need to be listening for the whistle that called them, loitering expectantly as they await the arrival of their daily rations.

ABOVE **Grass day, when
the cows are moved to the
allotments. Delicate heads
of cotton grass dance in
the gentle breeze.**
RIGHT **A covering of snow
at the beginning of May.**

The prospect of a lengthy uphill climb to the moor gate never seems to deter the children.

As with people, they all have their own unique character. Some are free-spirited and independent, and will break away to graze solo, while others can be found in the company of the same set of sheep. Cows tend to have a matriarch amongst the herd, but a head sheep is less common – although there will always be one who barges their way to the front of the queue.

The prospect of a lengthy uphill climb to the moor gate never seems to deter the children, even though they know full well that they may be burdened with carrying one of the younger lambs or have to constantly urge on a particularly belligerent yow. When weeks before we had to hurry, when time was of the essence and tempers frayed, there might have been mutterings of discontent. Not so now; even the sheepdogs appear to be fully aware that this is the end of their busiest time of year.

I recall one particularly glorious evening. After spending the day in the sheep pens, preparing the sheep for turning them back to their moorland home, there was only one more job to do.

We had sat down for tea, leaving the sheep quiet in the pens to mother up before we would set off to return them to their rightful place.

'Who's coming to the moor?'

On this occasion, Edith, who is a keen and able hand amongst the sheep, was the first to be heading out of the door.

'Wait up,' I called, 'you'll need this.'

I handed her a spare whistle, a homemade project carefully crafted from a stainless-steel teaspoon. It had amused Miles for a whole afternoon, filing and hammering until finally the whistle was pitch tested and deemed usable. It had since then lain on the mantelpiece awaiting Reuben to drill a hole in the stump that once was the teaspoon handle so that it could be attached to a lariat and worn around the neck.

We set off, Edith, Clemmy and Annas, myself and two dogs, Kate and the younger Taff. The sheep, hungry after a day in the pens being pestered by their suckling lambs, were now keen to fill their bellies. We opened the pen gates and they streamed forward and made away up the fields. Their initial sprint to freedom would soon slow to a manageable pace so for now we needed to hold the dogs back. Using the dogs too early would tire the sheep

and we would pay the price later when they refused to walk another step.

The children chattered amongst themselves as we walked but I knew that the conversation would soon dry up as the fields steepened and the flock slowed. The uphill hike would leave them breathless, and dogs would now be called upon to keep the dawdling sheep on track.

It would always take a joint concerted effort to get all the sheep to the gate, with differing mothering levels keeping us on our toes. Some, with lambs at foot, can become protective and turn on the dogs, or panic they have lost their lamb and make a run back to the gate. While for others, the idea of the freedom ahead overrides everything, and they are off, their lambs abandoned and struggling along behind. Annas and Clemmy were flagging and would occasionally flop onto the grass and lie quiet for a minute before setting off again. Edith and I could not wait, tapping the lambs with my crook to gee up those that lagged behind, whistling to the sheepdogs that were busy flanking the sheep at either side. It is easy to accidentally leave a yow behind, with so many slack spots and gutters in which a sheep can be hidden from view. There really is nothing worse than arriving at your destination, turning to survey the scene only to see a sheep in the distance, then having to retrace your steps in order to bring it to join the rest of the flock.

Edith was by now carrying a lamb, a younger one that had not been managing to keep up. Rather than constantly having to harass the poor little body, it was easier to just pop him under her arm for the last few hundred yards. He could be set down and reunited with his mother the moment that we got to the gate.

Finally, we had almost made it; the end was in sight. Leaving the children and dogs behind

the sheep I loped to the gate, opening it wide and propping it back. The front runners of the flock knew exactly where they were going, only pausing momentarily to check whether they had their lamb in tow. If all were present and correct they stepped through the gate onto familiar turf and were gone, spreading away wherever their fancy took them.

Finally the last yow went through the gate. Edith dutifully put the now bleating lamb down beside a clump of heather and retreated. For a moment the lamb looked around, dumbfounded and seemingly not sure what to make of his new surroundings, and then his bleat was answered by a yow striding purposefully back. She stopped short of coming all the way, eyeing up Edith and myself suspiciously. Another bleat to her son and heir and he made a beeline for her and was straight under her, suckling vigorously while she nuzzled his tail top.

'There,' I said, pleased with myself, the operation having gone smoothly. No casualties, no major hiccups, no complaints . . . Well not that I could hear, anyway. I could see Annas and Clemmy chatting away, but they were still too far away for me to be able to hear the conversation. Kate and Taff, having realized that their work was done, contentedly nosed around in the heather. I was impressed with Edith; she had kept up and now, resting her chin on my crook watching the flock disperse, looked very much the archetypal shepherd of the hills.

They say that you do not plant a tree for yourself but for your children, and the rate of growth here in the hills is pitifully slow.

Walking back through the meadows this month we keep to the edges. From now on we will avoid trampling the meadow grass whenever possible but skirting the edges means we notice and pick up any topstones that have become dislodged from the drystone walls. We pass plantations which are relatively new additions to Ravenseat. Over the last twenty years we have fenced off many strips of land that border the fields and have little value in terms of grazing. Steep, wet or just impossibly rocky, now they grow trees. They say that you do not plant a tree for yourself but for your children, and the rate of growth here in the hills is pitifully slow, but already we are able to forage amongst the shrubs and saplings that scratch a living in our poor acidic heavy soil. An ancient lone crab apple tree stands stark against the gable end of the Far Ings

Crab apple blossom on the tree by Far Ings barn.

barn. Unable to grow beyond the confines of its sheltered position, it will never elevate itself above the rooftop but every year without fail treats us to a sharp shock of pink before gifting us with the smallest, hardest, bitterest fruit you could wish for.

Lone hawthorns grow in exposed, rocky areas, stunted-looking, like Yorkshire's answer to bonsai trees. The subtle scent of the blossom is a joy to behold, a fragrance that is as dainty and delicate as the exquisite tiny flowers themselves. There is little point in picking the blossoms as their beauty is short lived and the petals soon wilt. Far better to leave them in situ and await the fruit in the autumn when the ripe haws will make a decent jelly, a perfect astringent accompaniment to darker stronger meats.

Mountain ash is flowering now too, creamy white flowers looking much prettier than they smell. Also known as rowan, this is one of the few trees hardy enough to withstand the extreme climatic condition of this altitude. The trees are squat, often gnarled, a tangled knot of thorny branches all askew, bending as though stooping beneath the biting easterly wind. Long associated with witchcraft and believed to hold magical properties, the wood would be used to make various small dairy utensils in the seventeenth and eighteenth centuries, even, I believe, into the early nineteenth century, and stirring the milk with a wooden spoon would surely ward off the evil spirits that were blamed when the butter refused to 'come'. Rowan Tree Ghyll nearby recognizes the importance of the ancient tree growing there, both as a marker and as a place where this wood could have been harvested. This spindly tree was also said to protect anyone taking refuge beneath its branches from the likes of the devil, though it's hard to see who could really huddle in such a small space.

Little of our farm is blessed with ancient native woodland, timber being a commodity that was in high demand during the industrial periods of the seventeenth and early eighteenth centuries when lead and coal mining in the Dales were at their peak. Smelting required a great quantity of wood to fuel the furnaces and much of what remained of the local forested areas was felled.

Fortunately a few pockets can still be found. The inaccessibility of the steep-sided gorge below the pastures through which Whitsundale beck runs lends itself perfectly to the preservation of a small oasis. A secret fairy dell where you can lose yourself amongst the tinted symphony of thick bowering trees under which we will stand, necks craned, staring upwards through the criss-cross of branches and to the sky. It would be no exaggeration to say that as inhabitants of open, windswept moorland places, this sylvan setting is as intriguing and remarkable to us as a foreign land.

Here leaf litter lies undisturbed, rotting slowly, creating a rich, fertile, but (because of the incline), shallow soil. A range of damp- and shade-loving

plants cling perilously to the steep banks. From within the deep fissures of the rock face protrude the star-shaped fleshy leaves of common butterwort. Usually it is the flower head of the plant that attracts attention, but although dainty and intensely purple, it is the fragile, long, furred stem that begs closer inspection. A carnivorous plant, it feeds upon midges and small flies trapped in the sticky fluid secreted by the stem.

In this shady, wooded enclave no breezes blow and an eerie stillness hangs in the air. Rivulets of iron-rich peaty water feed into the gently flowing river that picks up speed as it rounds the corner and funnels into a fast-flowing channel and over the bed of shelving rock, from where it cascades into a foaming dark pool. Jagged shards of petrified mosses, dislodged from their precarious foothold above, lie beneath the sheer rock face. Porous and brittle, they resemble pumice, their tiny honeycombed cavities home to a range of crawling insects. Strewn haphazardly beside the water's edge are enormous upturned boulders and ragged rock faces, timely reminders of flash floods and downpours upstream that have the power to shift

RIGHT **Boggle Hole when the bluebells are in flower.**
BELOW LEFT **One of the water babies.**
BELOW RIGHT **Violet gathering up wild garlic.**

the seemingly immovable. We can only venture into this place in a droughty spell when the water is at its lowest level or at least warm enough that we can wade through the deeper dubs.

The children scramble along the soft banks, picking their way over moss-covered stones. Dressed in swimsuits and pumps, with one hand they grasp at the clumps of vegetation in an attempt to steady themselves. The snapping of stems releases the pungent aroma of ramsons, also called wild garlic. The broad leaves and white lacy flowers grow in abundance in this secluded quarter

and we'll take the opportunity to bring them to the farmhouse kitchen. They are milder in taste than you would expect, so plenty are required to pack any kind of culinary punch, but the subtle garlic flavour is fantastic. All parts of the plant are edible but its flavour depletes during cooking so it's best to add it towards the end. The plant is purported to have antiseptic, antibacterial and antibiotic qualities. Not that I am saying that it is a cure-all, but for the thrifty amongst us it is an easy, cheap way to liven up omelettes, prettify soups and flavour savoury dishes.

Wild Garlic Lamb with Hasselback New Potatoes

Wild garlic has a more subtle flavour than the bulbs, and is less faff to prepare than fiddly cloves. We only eat meat from older lambs.

Prep time 20 minutes, plus marinating overnight / Cooking time 1 hour 35 minutes / Serves 4

INGREDIENTS

1.2kg lamb shoulder joint

30g wild garlic, washed

12g bunch mint

3 tbsp Yorkshire rapeseed oil

zest of 1 small lemon

1 tbsp Dijon mustard

Hasselback potatoes

800g new potatoes, scrubbed

40g butter

1 tbsp Yorkshire rapeseed oil

2 sprigs rosemary

salt and black pepper

Tip

Use the meat juices to make a gravy and try serving with spring greens.

METHOD

1. Place the lamb in a roasting tin and using a sharp knife, make a few incisions into the surface then season with salt and black pepper.

2. Chop the wild garlic and mint leaves finely, then mix in a bowl with the rapeseed oil, lemon zest and Dijon mustard. Spread the herb mixture over the lamb, then cover and transfer to the fridge for the flavours to develop. Leave overnight if you have time.

3. Remove the lamb from the fridge about half an hour before you are going to cook. Preheat the oven to 190°C/170°C fan/gas 5. Roast the lamb on the middle shelf of the oven for 1 hour 20 minutes for medium-cooked lamb.

4. Meanwhile, cut slits in the potatoes at 3mm intervals, leaving the potatoes intact and connected at the bottom. The best way to do this is to place each potato on a wooden spoon which will prevent you from cutting all of the way through.

5. When the lamb has about half an hour left to cook, put the butter and rapeseed oil into a medium-sized roasting tin and place on the bottom shelf of the oven for 2–3 minutes, until the butter has melted.

6. Next, pick the rosemary leaves from their stalks and finely chop; discard the stalks. When the butter has melted, sprinkle over the chopped rosemary, then add the potatoes, season with salt and pepper and toss gently to coat the potatoes in the butter and oil. Arrange the potatoes cut side up in the roasting tin.

7. Return the roasting tin to the bottom shelf of the oven and roast for 30 minutes. When the lamb is ready, take it out of the oven, cover with foil and leave to rest for 15 minutes. Turn the oven up to 200°C/180°C fan/gas 6 and baste the potatoes with the butter and oil, then place the roasting tin onto the middle shelf of the oven until the potatoes are tender and the skin is golden.

8. Carve the lamb and serve with the hasselback potatoes alongside.

Chocolate Orange Chelsea Buns

This is a simple recipe that the kids love making, but the buns still look and taste great. They come out of the tin looking like a tear-and-share creation – with many small hands ready to pull it apart.

Prep time 20 minutes, plus 1 hour 40 minutes to prove / Cooking time 25 minutes / Makes 12

INGREDIENTS

Dough

500g strong bread flour,
    plus extra for kneading

7g fast-acting dried yeast

50g caster sugar

1 tsp fine sea salt

zest of 1 small orange, juice
    reserved for icing

50g butter, melted

200ml warm milk

1 medium egg, beaten

oil for greasing

Filling

50g soft butter, plus extra
    for greasing

75g light brown sugar

2 tsp ground cinnamon

100g sultanas

100g dark chocolate chips

Topping

50g apricot jam

2 tbsp boiling water

75g icing sugar

1½ tbsp orange juice

30g dark chocolate, melted

METHOD

1. Sift the flour into a bowl, then stir in the yeast, caster sugar, salt and half the orange zest.

2. Mix the melted butter, warm milk and egg in a jug. Make a well in the centre of the flour, then pour the liquid into the bowl and bring the mixture together with a spatula. Turn the dough out onto a lightly floured surface and knead by hand for 10 minutes, until the dough is smooth (or 5 minutes if using a mixer fitted with a dough hook). Return the dough to a lightly oiled clean bowl, cover with a piece of oiled cling film or a clean damp tea towel and leave in a warm place for an hour until the dough has doubled in size. Grease and line a 20 x 30 x 5cm deep baking tin with baking parchment.

3. Knead the dough again for a couple of minutes on a lightly floured surface, then roll out to a rectangle approximately 35 x 45cm. Spread the soft butter evenly over the dough, leaving a 1cm border at one of the long ends. Mix the sugar, cinnamon, sultanas, chocolate chips and remaining orange zest in a small bowl, then scatter evenly over the dough.

4. Dampen the clean border of the dough with a little water then roll up towards it, pinching the seam closed. Cut the roll into twelve even slices. Place the slices cut side up into the prepared tin, spaced about 1.5cm apart. Cover the tin with oiled cling film or a clean damp tea towel and leave to prove for about 40 minutes, until doubled in size. Preheat the oven to 180°C/160°C fan/gas 4.

5. Bake the Chelsea buns for about 25 minutes, until golden and well risen. Allow the buns to cool slightly in the tin before transferring to a cooling rack.

6. Warm the apricot jam in a small pan with the boiling water, then brush over the top and sides of the Chelsea buns. To make the glaze icing, sift the icing sugar into a bowl and stir in the orange juice to make a smooth icing.

7. Drizzle the chocolate over the buns, let it set slightly, then drizzle over the icing.

June

June for me is the pinnacle of the seasons at Ravenseat. While May was a gradual awakening, the advent of June throws it into full flush. Every which way you look, there's new life unfolding, bright and beautiful, brimming with colour, life and energy. The meadows come alive now the sheep have moved on, and work their way through a veritable rainbow. First the yellows appear, fields of shimmering gold, filled with globe flowers, buttercups and marsh marigolds, their heads bobbing in the wind.

The common buttercup, this joyous little sunshine plant, is popular with insects, though little else. Poisonous in its entirety to grazing animals, it is unpalatable and steadfastly avoided by all but the hungriest, affecting their digestive system to varying degrees. It only becomes edible when preserved as hay or silage and is an indicator species, signalling that the soil here is extremely poor, heavy and acidic. Within days the pinks start to sneak in – fragrant orchids of the spotted and marsh varieties, and the delicate, star-like ragged robin – before the deep indigo and vibrant purples of devils-bit scabious complete a symphony of botanical colours that makes your heart sing. To fully appreciate this spectacle you must get down to ground level – grass roots, I should say – lie in the sward and on the drier banks, and look around at the tiny yellow and purple-flecked mountain pansies, listening to the hum of insects and the chatter of birds.

There is a constant battle to fend off the encroaching rushes, clumps of which proliferate in the permanent pastures. They are welcome in the areas of permanent pasture as here they act as cover, a windbreak providing protection for birds and mammals. Within the midst of a slightly elevated tussock you might chance upon a neat clutch of snipe eggs. Hares too will take advantage of the dense clusters of stems, bending the stalks to form a simple arch, where they can lie low, out of view and beneath the wind, and sit tight until any danger has passed.

Early nesting birds such as curlews and lapwings have their chicks by now, and the chatter of the parents is a constant accompaniment to our days, as they attempt to divert our attention from their young.

Curlews are instantly and easily identifiable, distinctive thanks to their incredibly long, thin, down-turned beaks, perfect for foraging for insects and worms. Their numbers have sadly been in decline in recent years, so birds that were once a familiar sight soaring high in the skies above farmland nationwide are now critically endangered.

Their dramatic, haunting calls echo far and wide, but they are difficult to observe close-up. A naturally nervy bird, they are notoriously coy when going about their activities and will take to the wing even when not disturbed.

Last year Sidney happened across a curlew nest in the Big Breas meadow containing four of their distinctive, large, mottled-beige eggs. It was a pleasing find and, wishing to find out more about the daily habits of this elusive creature, we rigged up a remote camera trap. It was a real privilege to have a bird's-eye view – literally – of an event that was common on our doorstep but which we never got to see. I started to feel a personal connection as I became familiar with the curlews' habits, and it became part of the daily routine to check on them when out in the field, and via the camera later in the day. I would commando-crawl through the grass to get a closer look without disturbing the family.

After a week of the mother and father taking it in turns to sit on the eggs while the other went off in search of food, another, slightly smaller egg of lighter hue inexplicably appeared in the nest. How had that extra one appeared? I researched the possibilities and discovered that five eggs is almost unheard of from one set of curlew parents. The most likely explanation was that, rather like a cuckoo, another female curlew, not taken with the idea of impending parenthood, will sometimes find another nest and lay an egg there, leaving the other parents none the wiser that they are partaking in an adoption process.

ABOVE **Raven amongst the globeflowers.**
OPPOSITE **A curlew.**

After four weeks, the original four eggs began to crack, and little brown-and-yellow bundles of fluff emerged, their distinctive beaks already clearly indicating their species. We watched via the camera, in awe.

Inevitably the fifth egg was still lying there, unmoving, not as far down the line in development as the others. But the adult curlews were now busy with their new chicks.

'What do we do?' I said to Clive, wanting the poor abandoned chick to have a chance of hatching.

'Bring it back and put it under one of the chickens?' he suggested.

A great idea! So off I went, keen to help, but already wondering about how I'd feed it. The hen's mash wouldn't be suitable; somehow I'd need to find a way to catch live insects.

As it happened, the egg remained cold and lifeless, despite the broody hen's best efforts. In reality it was probably a blessing. Having a hand-reared curlew to care for might have been a step too far, even for me.

The children were fascinated by the new curlew family, although the birds' endangered status didn't sway them – in their world the curlew was a common visitor, and all wildlife at Ravenseat holds the same allure. They are forever returning to the farmhouse with an animal under one arm, or a bird held out like a prize find, although I try to encourage them to leave the animals out in nature and just enjoy watching, unless absolutely necessary. While I appreciate their desire to do right by the creature, we already have enough animals to take care of each day.

My heart sinks when they turn up with a bird, a broken wing hanging limply from its frail body, as, truth be told, there is no hope for recovery. Birds are notoriously bad patients, often succumbing to stress long before the injury takes its toll.

We did successfully help to feed an owlet that we found beside the manger in the stable. There comes a point when young owls leave the nest, though they stay in relatively close proximity, their inability to fly properly preventing them from going too far away. At this stage, known as 'branching', they are still fed by the parents and are vocal, issuing unusual begging calls during the early evenings. This owlet, a tawny, was thin and in poor condition, so after consultation with our vet we began a nightly regime of top-up feeds – diced-up raw chicken and a glucose drink given via syringe. The owlet, wrapped in a tea towel, would swallow heartily; then we would carefully replace him in the stable, assuming that he was

ABOVE **Feeding the barn owl.**
BELOW **A hare.**
OPPOSITE **A curlew chick.**

still being tended by his parents but that for some reason he was not thriving as he should. After a week, he disappeared. You never really know what happens in these cases; you just do your best and hope that your efforts were not in vain.

Foxes and badgers aren't a common sight at Ravenseat, as open country and moorland don't lend themselves to the cover they look for. They prefer to make their homes in hedgerows and woodlands. The drystone walls, though, do provide a haven for mice, and lifting a rock in late springtime will often expose a moss-filled nest of baby field mice hiding underneath, their eyes still tightly closed, hairless bodies cuddling together for warmth.

Down at the river, away from the main current, inlets form in the calm, clear waters and make a perfect home for the smaller fish and tadpoles. The children swim in amongst them, fascinated by the wriggling black apostrophes tickling against their bare arms and legs.

June is the one month where there is a slight hiatus at Ravenseat. The pressure is off, if only for a short while, and the jobs don't feel quite as urgent. There is a sense of respite, and if there was ever a time to go on holiday it would be now – but of course we don't! Instead we see it as a time to take in the beauty of our surroundings, which passes us by in busier times. We picnic under the beautiful hazy blue skies, spend a few hours paddling in the river or relaxing on the 'beach', a sandbank that has formed on one of the river bends. I love to get out on the horses when I can, although I find it hard to ride out simply for pleasure. It seems indulgent, somehow; therefore I will combine business with pleasure and task myself with going to check on or move the stock.

Boggle Hole, accessible only when the water is low.

Most people see summer as July and August, but up here – and for most people who farm in the uplands – it is June. Nothing can beat the vibrant, summer feel of the sixth month.

Most people see summer as July and August, but up here – and for most people who farm in the uplands – it is June. Nothing can beat the vibrant, summer feel of the sixth month.

There are of course still some animals that need daily attention. Yows with twins or those that need some extra care are in the allotments and high pastures, a kind of halfway house before they can join the rest of the flock on the moor. They are torn between a desire to get up to the moorlands, where they feel most at home, and down into the developing meadows, where they have spied the lush grass growing, a perfect meal for every gluttonous sheep.

The cows too have long forgotten their months in the barns and are having the time of their lives roaming freely. The land caters for all their needs.

We head into the fields and check the walls and boundaries. Keeping on top of drystone-wall maintenance is a must, if the dreaded clank of machinery grinding over a stone at haymaking time is to be avoided. No repair is more frustrating or needless. It is also time for the barns and sheds to get a good clean-out, and the midden – the muck heap – is topped up with old bedding and dung. There is finally some spare time to clear up after the winter, and prepare for the months ahead.

Nettlebeds appear, growing in profusion in places untouched by the mower, on the rocky

foundations where sheep folds once stood and in quarried places where spoil remains and upturned stones protrude. Their worth is considerable, as a habitat for insects, ladybirds and moths, and a haven for butterflies.

It is well known that nettles are edible for humans too, and the newly grown, bright green tender stems and leaves can be used in recipes as a poor man's substitute for spinach or wilted greens. I'm told they are good for inflammation in humans, and can help any number of ailments, from tackling hay fever to lowering blood pressure. Violet and Edith spend many happy hours gathering the nettles, no doubt exhilarated by the element of

danger that comes with tentatively snipping the stinging tops, one child armed with haberdashery scissors, the other with gloves, and placing them into a basket while trying to avoid the customary itchy weals. Every year there are in-depth conversations on the best way to grab a nettle without being stung. They are grasping the nettle, literally and figuratively.

I add nettles to any meal as a replacement for spinach, and they are particularly delightful simply blanched and served with a fresh poached egg. While they don't have the strongest of flavours, there is a certain satisfaction in eating something for nothing, that the children have harvested. On one occasion Raven made nettle tea, which turned out to be a thoroughly joyless experience never to be repeated, although surprisingly the oddly bottle-green-coloured nettle hair rinse was a remarkable success. It wasn't the most fragrant of potions but the girls' hair shone.

The dark blue of bilberries can be spotted around the beck side and precarious overhangs. Growing in clumps, the best hauls are under the leaves, where you might find a few decent plump berries the birds haven't yet spotted. The kids clamber to get them, balancing on each other's shoulders, their faces contorted as they nibble on the tiny, super-sour morsels.

The chance of including bilberries in a recipe is slim. The children spend an afternoon picking them, but rarely have anything to show for their efforts, returning home empty-handed with

an activity something akin to a poor man's truffle hunt. Pignuts' whereabouts are betrayed by the feathery-looking plant with small white flowers that grows above them, resembling cow parsley. Rife around Ravenseat, when gently pulled from the ground a pignut resembles a hazelnut with a papery brown skin that rubs off between your fingers. The crunchy texture and nutty, crisp taste is halfway between a water chestnut and a radish, and the children collect handfuls of them, filling large enamel cups and devouring them like sweets. Competition is rife as to who can find the biggest one. They will kindly offer one to anyone who passes through the farm, although unsurprisingly the soily little offering doesn't get many takers.

purple-stained faces and fingers. Like the wild strawberries that occasionally are spotted near The Firs, there's no point putting them in your pockets for later and squashing them; the pleasure of eating the fruit is better enjoyed then and there.

The children are also partial to a few of the edible flowers – lady's smock is a favourite, with its pink-and-white flower. Regularly scattered across plates in fancy restaurants, it is merely there to prettify a dish. A hint of pepperiness is all you get as you consume it, but for the children, it is all part of the novelty of eating homegrown produce.

Another favourite in June is pignuts – or *Conopodium majus* to use their scientific name. The children get stuck in, digging into the soil, and coming up triumphant with the small white prizes,

Yorkshire Pan Bagnat

A showstopper sandwich, this is a great showcase for local produce and makes for a picnic in itself. The bread from the middle can be used in the Tomato, Chilli and Spinach Panzanella Salad on page 153, otherwise the chickens are happy to oblige!

Prep time 15 minutes, plus up to 4 hours chilling time / Serves 4–6

INGREDIENTS

1 white boule loaf
   (approximately 20cm in
   diameter)

2 tbsp Yorkshire rapeseed oil

60g rocket

2 thick slices of cooked
   Yorkshire ham

4 tbsp caramelized onion
   chutney

100g Wensleydale cheese,
   sliced or crumbled

2 vine tomatoes, sliced

3 medium eggs, hard-boiled
   and sliced

salt and black pepper

METHOD

1. Slice a lid off the top of the loaf, then scoop the bread out to hollow out the loaf. Brush the inside of the loaf and the lid with the oil.

2. Start to build the pan bagnat by laying half the rocket into the base, then add the ham, followed by the chutney. Next, add a layer of cheese, followed by the tomatoes and eggs, then the remaining rocket and a pinch of salt and black pepper.

3. Wrap the loaf tightly in cling film, place on a small tray and put a weight on top to press the filling down. Transfer to the fridge for at least an hour, or longer if you have the time (4 hours maximum).

4. Cut the pan bagnat into wedges and serve with a salad.

Tip
Use the scooped-out bread to make fresh breadcrumbs.

Elderflower and Summer Berry Pavlova

Elderflowers are a neglected hedgerow gem, and give a lovely flavour to this dessert. Adding the edible flowers that the children will also have collected is a sure-fire way to guarantee they ask for seconds!

Prep time 15 minutes / Cooking time 1 hour, plus 1 hour with heat off / Serves 6–8

INGREDIENTS

4 medium egg whites

225g caster sugar

1 tsp cornflour

½ tsp white wine vinegar

400g mixed berries:
strawberries, blackberries,
blueberries, raspberries

3 tbsp elderflower cordial

300ml whipping cream

handful of edible flowers

METHOD

1. Preheat the oven to 150°C/130°C fan/gas 2. Draw a 25cm circle on a sheet of baking parchment, using a dinner plate as a template.

2. Place the egg whites in a clean bowl and whisk with an electric mixer until they form stiff peaks. Add the sugar, a tablespoon at a time, whisking constantly until all the sugar has been added. The meringue mixture should be thick and glossy. Gently fold in the cornflour and white wine vinegar.

3. Place the baking parchment circle on a large baking tray, with the drawn template face down, and secure with a small amount of meringue in each corner. Spoon the meringue onto the circle and gently spread using a spatula, then make a large indent in the centre.

4. Place the baking tray on the middle shelf of the oven and bake for an hour, until the meringue is crisp on the outside. Turn the oven off and leave the meringue in the oven for a further hour.

5. Remove the meringue from the oven and allow to cool completely, then gently remove from the parchment and place onto a serving plate.

6. Pop the fruit into a bowl and sprinkle over 1 tablespoon of the elderflower cordial.

7. Meanwhile, pour the cream into a bowl with the remaining elderflower cordial and whisk to form soft peaks, using an electric mixer.

8. Pile the cream into the centre of the meringue and top with the soaked fruits. To finish, arrange the edible flowers on top of the pavlova to decorate.

Tip
Try adding some chopped roasted hazelnuts to the meringue before cooking.

July

Make hay while the sun shines. Oh how straightforward that sounds . . . But it is not so simple when living on a hill farm with a short growing season and predominantly high rainfall. Our obsession with the weather becomes all-consuming during July, as the two big tasks for the month – haymaking and sheep shearing – are urgent, time-sensitive and entirely at the mercy of the elements. And so as June draws to a close, and with a little rest and relaxation under our belts, we are preparing for a month of hard physical labour. It is, in essence, time to reap what we have sown in the form of hay and wool. If the month goes to plan, the heady scent of freshly mown hay will fill the air.

Once upon a time hay 'got' in June was something to really talk about, a novelty. Nowadays a field of grass might be cropped repeatedly from early spring through to early autumn. It can be short cropped for silage, often to feed 'zero-grazed' animals, those kept in a farming system whereby they stay within the confines of the buildings. Not so in the hills. Here, where the winters are long and the summers short, there will be only one opportunity to mow for hay.

The first job when tackling this annual task is getting the rusty machinery out of the nettlebed where it was unceremoniously dumped the previous year. With any luck our resident mechanical expert Reuben will be on hand to tackle the loosening of seized bolts and gears, and get us moving. Then it is time to cut down those heavenly meadows, allow them to dry and turn them into hay bales.

Despite the heat, throwing bales into a trailer is not a task to be undertaken while wearing skimpy outfits, or skin will soon be covered in itchy weals, scratches and bites. It's a cover-up job, only tolerable in trousers and long-sleeved tops. The twine that binds the bales will cut deeply into any hands that are as yet uncalloused. Blisters from using the hay rake are commonplace too, as is the agony of minute splinters from dried thistles and thorns. All this means that gloves are a must.

A relatively small area of our total land mass is hay-timed – only fifty acres is really suitable. With so much of our land being steep or boggy, great swathes are left untouched. Where once an army of men with scythes would have set to, cutting every last blade of grass that they could reach, we concentrate our efforts on areas where we can safely get the tractor and mower. And now we are more mindful of leaving space for nature too, buffer zones where hares, rabbits and field mice can reside.

There is something bittersweet in cutting down the meadows that have provided so much beauty through the previous month, although crucially by now the flowers and grasses have all gone to seed and the ground-nesting birds all fledged. Hay in the barn is money in the bank, and it feels like our insurance policy. We are making preparations for winter, and come hell or high water the animals will be well fed. So moving the bales into the barn, breathing in the summer smell of grass, herbs and flowers, leaves us with a sense of overriding satisfaction.

If the weather has been catchy, with no clear stretch of the hot, dry conditions required, the task of haymaking has been known to creep into August, and even the first week of September. One might imagine this is of no consequence as long as it's a good crop, but nutritionally the grass is losing its goodness every day onwards from its peak in July. We need to get our crop in at its very best, if possible, as we rely on it to carry the sheep and cows through the leaner times.

PREVIOUS PAGE **Annas** raking the hay.
ABOVE **Our steep** and rugged terrain is best suited to a small conventional baler.

There is something bittersweet in cutting down the meadows that have provided so much beauty through the previous month.

Then it is time for
a different smell to
dominate the air, the
musky, oily fragrance
of freshly shorn wool
indicating that it is
sheep-shearing time.

Then it is time for a different smell to dominate the air, the musky, oily fragrance of freshly shorn wool indicating that it is sheep-shearing time. This is also ideally carried out in dry weather, partly as the sheep are far more pleasant to handle, but, more importantly, because stacking and storing wet wool is to be avoided, for it will rot.

The sheep are gathered down from their heafs at the moor and brought to the fields around the house to await their turn. The hoggs (young sheep who haven't yet been shorn) go first as, never having been clipped, they are carrying fifteen months of growth. They will never produce a fleece as plentiful or of such high quality as they do in that first clip, and there are about 200 to 250 of them to tackle. It can be a baptism of fire as they are hardest to clip, with a profusion of wool everywhere; we must avoid nicks to their tiny teats, not yet developed by motherhood. As flight animals, they are understandably nervous at being held, but a skilful shearer will keep them constantly moving with small, almost imperceptible shifts. It relaxes the sheep a little to feel movement rather than restraint, calming their natural instinct to try to escape.

Despite their reservations, it is a relief for the hoggs when the thick fleece is gone. Depending on the weather, it will be starting to make them hot and sweaty, and they seek out drystone walls and rocky outcrops amongst the peat to have a good scratch.

Luckily, itchy, hot wool is as bad as it gets, as thanks to the height of Ravenseat, and lack of tree cover, we don't tend to be bothered by blowfly strike – every shepherd's nightmare.

Shearing is my favourite time of year. There is a basic satisfaction to the sheep coming in, heavy with wool, and going out lighter and neater, minutes later. I love the physicality of it, and seeing my technique speed up and improve with each sheep.

The children have all tried their hand at it, but I've never taught them properly, as the 'how' is an ongoing dilemma. Clive and I both cut the old-fashioned way, learnt when people graduated from hand-shears to electric. We place the yow on her bottom, and clip while standing, starting at the neck and down one side, then the other.

The modern method is called the Bowen technique, and the yow is laid on her back, with the shearer going in at the hind leg, and 'shaving' (sorry) a few seconds off the time it takes.

I can't decide whether to teach the children the way I clip, or send them on a course to learn

OPPOSITE **The Anty Johns flock, newly shorn.**

the new way, and somehow neither have happened. For Miles in particular, though, who has his heart set on being a farmer, it is high time we made a decision.

The acidic soil of Ravenseat is unsuitable for most crops – we are no Garden of Eden – but we do make sporadic attempts to grow vegetables. Previously the results have included a few meagre salad leaves, potatoes and a cucumber plant that looked, frankly, obscene, producing a single, small, stunted cucumber with a copious covering of fuzzy down. But last year the attempt was begun in greater earnest by Violet, Miles and Sidney. There was nothing as organized as set vegetable patches and raised planters, but the mission was on – to hunt out the most fertile soil, wherever that might be, whether above the septic tank or in the area known as the 'graveyard' (beside the woodshed that was once a chapel), or above the septic tank!

Radishes were our first vegetable of choice last summer and, ignoring the instructions to 'sow thinly', the contents of the packets were scattered liberally. It turned out the poor soil was no deterrent. Somehow they not only survived but flourished, and it soon felt as though we had too many radishes to contend with. Every salad became an excuse to use up the humongous crop, and when that novelty wore off, we began eating them like sweets. Eventually the penny lost its shine and the children, although initially enthusiastic about eating homegrown produce, stopped

Weeks later we found forgotten radishes still growing, reaching sizes that would rival most turnips.

harvesting them. As a result, weeks later we found forgotten radishes still growing, reaching sizes that would rival most turnips.

We did well enough with tomatoes – perhaps because they were planted indoors in the sunroom, which really captures the warmth – plus the spinach and chillies, although the cauliflowers failed.

We also tried our hand at potatoes and carrots. Unfortunately these require a certain amount of patience, something the children were lacking, and they kept pulling them up to 'check if they were ready yet'. This meant we ended up with what could be passed off as new potatoes and micro carrots in a high-end restaurant, but in reality were underdeveloped midget veg that would clearly have benefited from another few months in the ground.

I could not blame them, as there are areas where my own patience is seriously lacking, particularly with anything that takes work once it is out of the ground, prior to cooking. I feel that planting, tending and pulling should be the completion of my end of the bargain. After that I just want to use the vegetables. I have scant time for any tying and drying, as with onions.

Creamy Chicken and Roast Radish Casserole

This is such a lovely, colourful summer dish that makes the most of the vegetables I have (hopefully) managed to grow. Apart from asparagus, of course – that one isn't so fond of the Ravenseat climate!

Prep time 15 minutes / Cooking time 50 minutes / Serves 4

INGREDIENTS

15g butter

1 tbsp Yorkshire rapeseed oil

1kg chicken thigh fillets, skin on

2 banana shallots, finely chopped

100g pancetta, diced

1 clove garlic, crushed

20g plain flour

300ml chicken stock

14 radishes, trimmed and halved

150g asparagus, cut into 3cm
    pieces

5g tarragon, leaves finely chopped

100ml crème fraiche

100g fresh peas

250g broad beans

salt and black pepper

METHOD

1. Preheat the oven to 200°C/180°C fan/gas 6.

2. Gently melt the butter and ½ tablespoon of the oil in a large casserole pan. Season the chicken thighs, add to the pan and cook on both sides for 4 minutes per side until the chicken is browned. Remove the chicken from the pan with a slotted spoon.

3. Add the shallots to the pan with the pancetta and fry for about 5 minutes, until the shallots have softened. Add the garlic and cook gently for a minute.

4. Sprinkle over the flour, lower the heat and cook for a minute, then add the chicken stock, bring to the boil and return the chicken to the pan, skin-side up. Cover, reduce the heat and simmer for 30 minutes.

5. Meanwhile, put the radishes onto a small baking tray with the remaining oil. Place on the top shelf of the oven and roast for 12–15 minutes, until tender. Add the asparagus to the baking tray when there are 8 minutes remaining.

6. Remove the lid from the casserole and simmer for 5 minutes, uncovered, until the sauce reduces slightly. Take the pan off the heat and stir in the tarragon and crème fraiche, then add the radishes and asparagus, along with the peas and broad beans. Simmer gently for 5 minutes and check the seasoning before serving.

Tip
Serve the casserole with some creamy mashed potato.

Tomato, Chilli and Spinach Panzanella Salad

A fresh, colourful salad, the bread does a great job of mopping up the sauce and making this a more filling dish.

Prep time 10 minutes / Cooking time 20 minutes / Serves 4

INGREDIENTS

1 medium red onion, cut into
    8 wedges

2 cloves garlic, crushed

1 medium red pepper, sliced

1 medium yellow or orange
    pepper, sliced

1 red chilli, deseeded and diced

2 tsp dried oregano

3 tbsp olive oil

1 small ciabatta loaf, torn into bite-
    sized pieces

100g baby spinach

400g heritage cherry tomatoes,
    halved

4 vine tomatoes, roughly chopped

80g black olives

1 small bunch basil, roughly
    shredded

salt and black pepper

Dressing

1 clove garlic, crushed

1 tsp Dijon mustard

3 tbsp olive oil

2 tbsp white wine vinegar

METHOD

1. Preheat the oven to 200°C/180°C fan/gas 6.

2. Put the red onion onto a large baking tray along with the garlic, peppers, chilli and oregano. Drizzle over half of the oil and season with salt and black pepper. Place the tray on the middle shelf of the oven and roast for 20 minutes.

3. Meanwhile, spread the chunks of bread out on a baking tray and toss with the remaining oil. Bake for 10–12 minutes on the top shelf of the oven, until lightly toasted.

4. When the vegetables are roasted, transfer to a bowl and stir in the spinach; it will wilt slightly from the heat of the vegetables.

5. To make the dressing, add all the ingredients to a small bowl and whisk to combine. Season to taste.

6. Mix the cherry tomatoes, vine tomatoes, olives and basil in a bowl and season with black pepper.

7. To assemble, put the warm roasted vegetables on the base of a large platter, then arrange the tomatoes on top, followed by the crispy bread. Drizzle over the dressing just before serving.

Tip

The salad is perfect served with roast chicken or some cold cooked meat as a buffet or picnic-style meal.

August

The summer equinox in late July comes all too soon, and if we have managed to get the wool crop in the bag and the hay barns full without major mishap we will feel a degree of thankfulness. In a perfect world we would have a little gentle summer rain come August, for now the emphasis is on replenishing the land with goodness; to see a freshening of what now seems bleached and bare, and a resurgence of greenery.

We give a helping hand to the 'fog' – this bright, nutritious regrowth – by heading to the muck midden and spreading the rotted-down muck as our natural fertilizer. Accessing the meadows with farm machinery is a precarious undertaking at the best of times, but if we have successfully navigated a field with the hay baler then it figures that we should be able to do the same with the muck spreader. Ideally the gentle rain begins shortly after we have a light covering of manure on the land, washing the nutrients into the soil. The focus is not on how much hay we can make, but how good we can make it, and a decent crop of herb-rich meadow hay is only achievable by using a natural approach.

We do not, therefore, use chemical fertilizers, herbicides or pesticides at Ravenseat. A perfect balance must be struck. Keep the right number and breed of animals over the right area and there's enough food to sustain them, while at the same time they provide a natural, organic fertilizer which helps everything to grow. It really is that simple – a system that has been practised and honed to perfection through centuries of farming in the hills. Regenerative, nature-friendly, sustainable agriculture is how it has always been at Ravenseat. Diets change; the climate has certainly changed; but the value of what places such as ours can deliver in terms of food production and environment are unquestionable, and if anything have become more valuable as time has gone on, as the onus shifts to long-term sustainability.

Meanwhile, in what we optimistically refer to as 'the walled garden', there can be found all kinds of plants that have stood the test of time, and which put down roots many moons ago. We have very little opportunity to tend this area, so it is testament to the hardiness of the gnarled blackcurrant shrubs and gooseberry bushes that they once again make it through each year. Not only do they have to survive the harsh climate and acidic soil, but marauding sheep and any amount of rabbits will inevitably wage war on accessible fruits, so life in the garden is more a game of survival of the fittest. Perhaps this explains the profusion of lethally sharp thorns on their branches.

If you can beat the birds to them, though, a crop of gooseberries is a guarantee. The sour, veiny, green berries make for the loveliest crumbles, taking over that role from the rhubarb that is now out of season. Acidic produce seems to fare better with the challenges of Ravenseat, while anything sweet faces real struggles.

I'd always wanted to learn more about how to use every plant Ravenseat has to offer and was happy to discover an old book of plant uses, from recipes to medicine. Reading up on the folklore surrounding many of the plants was fascinating. We all know that happening upon a clump of white heather while on the moor might bestow good luck upon the finder, but I had no clue that picking a bouquet of the delicate wood cranesbill

If you can beat the birds to them, though, a crop of gooseberries is a guarantee.

meant that an unseasonable deluge would ensue. I'm not generally one to be pulled in by super-stitions, but either way, we will err on the side of caution – woe betide anyone who plucks that flower before our hay is safely in!

We have recipes to try out for lady's mantle, great willowherb, sweet cicely, giant bellflower and purple and pink wood cranesbill.

The marsh marigold is one of the first edible wild plants of spring. It's commonly cooked as a pot-herb and, once upon a time, people welcomed it to their tables after a long winter of stored food – a spring tonic of sorts. The bright yellow flowers were used as an additive to poultry mash, dried and ground into a powder. I'm not sure whether or not this supplement was nutritionally helpful, but you can see how it would serve to brighten the yolks.

Inevitably there will still be sheep to clip in August, but we should have sheared the majority by now. Rather than rub our hands with glee at the bulging wool sheets, we must now consider what to do with the wool. Clive and I might manage to clip nearly 1,000 sheep over a week if the weather is good, and feel mightily pleased with ourselves, but the reality is that this once-valuable commodity is worth next to nothing. The only reason to remove it is for the welfare of the sheep; as a result of over 2,000 years of domestication, the sheep is now reliant on the annual shearing in order that it may thrive. It is a travesty that wool, as a natural, biodegradable product, cannot be put to good use in the textile industry – especially when there is a desire amongst shoppers to move away from plastics and man-made fibres towards more sustainable and renewable products.

It was in 2019 that we became involved in a pilot scheme to use our wool clip for an entirely new purpose. When looking across the moors it is plain to see that the ground is criss-crossed by regimented grips, ditches cut deeply into the ground to create a drainage system. Excess water is channelled out of the boggy areas and into these gutters, theoretically drying the land to make it easier to farm. Weather, and the flow of water, have seen them grow, worn wider and deeper into

the surface over time. The grips are now considered to be an entirely man-made disaster, and are devilishly dangerous to livestock that can drop into these steep-sided trenches and be unable to escape. I have even, albeit temporarily, lost children down them. Hidden as they are by the heather that grows over the top, it is easy to suddenly find yourself staring into a deep recess that, should you not be watching your step, will swallow you up. Some years ago, as a six- or seven-year-old, Miles completely disappeared from sight while we were out gathering the sheep. His muffled shouts meant that we had him fished out in no time at all, but still, it made him very wary for a good while afterwards.

Gripping became popular across the UK in the sixties and seventies to make the land more productive. The thinking was that if the land was drained, more animals could graze there, and therefore more food could be supplied. It made it harder for farmers to travel around the land and get across the grips, but they were encouraged to take part by the offer of government grants. Unfortunately there were unexpected side effects to these new trenches. They were impacting on flood management, with more water being carried away from the peat and down to the lowlands. Of even greater importance was the impact it was having on the environment in terms of the climate crisis. Moorland is a carbon sink and, acre for acre, is better at holding carbon than trees but, crucially, only if wet. Armed with this more up-to-date knowledge, it is clear the water table needs to be raised again, and there has been a huge push across the UK, and Europe, to try to rectify this.

All sorts of techniques are being put to use to block up the grips, such as dams of peat at regular intervals to slow the flow of water. Surveys were done by environmental agencies to see what could be done and how these ditches could be filled in without creating more issues. Using coconut and coir matting was a possibility, though ideally we wanted something easy to source, and preferably sustainable – so not really a product that needed to come from the other side of the world.

There was something satisfying about turning to nature for the solution, using our wool to give something back to the moorland which had provided a home for so many of the sheep, the farm able to solve its own issues internally. So, having been given the green light to put the plan into action, we worked on the coldest of winter days with expert guidance and practical support from the Peatland Restoration team. The wind lashed our faces; we worked our way along the grips with frozen fingertips, pushing the tightly rolled fleeces

down into the peaty channels to create semi-water-proof dams. It was an arduous but simple task. All that was required afterwards was for the digger that was on site to push the ridges at either side of the gutter inwards, to hopefully encourage the formation of fresh peat as the channels gradually silt up. What I wanted to avoid was too much wool being visible, as I envisaged numerous walkers coming down to the farm and reporting dead sheep strewn about the place.

The blocking of grips using wool is experimental, and in a trial phase, but nevertheless we are hopeful that it could be a valuable tool in the challenges that we currently face regarding flooding, erosion, carbon capture and climate change.

The enduring, timeless nature of Swaledale makes it a haven for tourists but, unlike so many tourist

Even after more than twenty years here, there are still new places for me to find on our land alone.

hotspots, the area still remains true to its roots. There is plenty of space to enjoy the natural unspoilt surroundings. It is a walkers' paradise, with footpaths taking in mile after mile of rugged moorland, rolling hills and pretty meadows. In any normal year we reckon to see around 16,000 walkers crossing our packhorse bridge on their way from Kirkby Stephen to Keld on the Coast to Coast footpath. The whole walk in its entirety is 192 miles from St Bees to Robin Hood's Bay, and Ravenseat is exactly halfway. These determined folk follow faithfully in the footsteps of Alfred Wainwright, who devised this particular walk; it usually takes them two weeks to complete. Every day from May right through until the end of September we would watch a slow trickle of walkers pick their way across the moor before speeding up on a short section of our metalled road. They brought business right to my doorstep, wanting nothing more complicated on most occasions than a cup of tea and slice of cake. It's fair

ABOVE Gathering and driving the sheep down Whitsundale.
BELOW A vole scurrying about its business.
NEXT PAGE Edith at the top of the High Force Falls.

to say that their presence was very much missed during the pandemic of 2020. Where once not a day would go by without there being some kind of conversation about the state of the weather, the path or their blistered feet, now there was nobody to be seen. Quiet upon quiet, extreme solitariness – never before and hopefully never again will we feel so remote and isolated, physically and mentally as well as geographically.

When the restrictions were finally lifted there came a new sense of freedom. After a year of confinement people were finally allowed to wander relatively freely again, and had a newfound appreciation for the countryside around us. The walkers returned to the hills, though there were fewer of the hardened long-distance adventurers who tended to be the ones with more experience, and far more daytrippers, many of who didn't realize

ABOVE **Midge at Alderson's Seat.**

the trouble they could get into on the moorland. The general assumption is that in England's green and pleasant land you're never going to lose your way or encounter any real danger, as you are always a stone's throw from the nearest town. There is no vast expanse of wilderness.

Maybe not, but the Yorkshire Moors are a surprisingly big area – 1,436 square kilometres, to be exact. Even after more than twenty years here, there are still new places for me to find on our land alone, and areas that I haven't been to or investigated. You only have to look at one of the drone videos that have emerged online in recent years to see the vast uninhabited sweeping expanses of the Yorkshire Moors. It is not a place to be caught out in the cold or dark, with nothing to help you find civilization or survive through the night.

The moorland can feel a desolate place at night – and not just for anyone lost on them, but for me too. We don't often go out there after hours, and even places you are familiar with become indistinguishable in the dark, and well-trod paths feel altered. On a cloudy night a smothering darkness envelops you and your confidence slips away, your senses heightened as your vulnerabilities are exposed.

But it is never silent. I was asked to do a radio recording for Jarvis Cocker's *Wireless Nights* on

BBC Radio 4. The request was simple: head out to the moor at the dead of night and record what you hear. The sensitivity of the audio recorder gave us an aural stream of squeaking, scuttling and scratching from the nocturnal inhabitants of these desolate hills.

We are so utterly reliant on technology, and in particular mobile phones, for direction and help these days, but few people realize there is no reception across huge swathes of the land around here. If you can't summon help via your phone, these two tips might prove invaluable:

- As long as your phone battery has some life, even without signal or an ability to ring out, the emergency services can trace your phone. If it is completely flat, they can't. If there is a chance people might be looking for you, preserve that last bit of power, as opposed to using the phone as a torch, no matter how dark it might be!

- Rather than constantly getting your phone out and turning the screen on to check if you have entered an area of reception, send yourself a text, leave your phone alone, and when a beep tells you the text has come through, you know you have reception. A simple idea, but it might just save you that extra bit of battery power – and from tripping over while constantly looking at your phone!

There is no part of me that would ever wish to discourage people from heading out into the countryside. The benefits, both physical and mental, are immeasurable. And of course, should the worst happen and an accident befall you in Swaledale then you can rest assured that you will not be left high and dry, as our local Mountain Rescue team are on call 24/7 to come to your aid and guide you to safety. They train at Ravenseat; it is the perfect place to put into practice some of their life-saving skills and hone their search and rescue techniques. The children are challenged to an enormous game of hide and seek: they are given a thirty-minute head start, and then it is the job of the rescue teams and search dogs to locate them. It might seem like just a bit of fun, and the children certainly see it that way, but it gives the rescue team the opportunity to work in a challenging and unforgiving environment – and, at the end of the day, enjoy tea and scones in the sunshine.

Yorkshire Sharing Board with Plum Chutney

Prep time 15 minutes / Cooking time 2 hours / Makes 2 x 500g jars / Sharing board serves 4

INGREDIENTS

Plum chutney

750g plums, halved, stoned
    and finely chopped

225g Bramley apples, peeled
    and chopped

200g shallots, finely chopped

200g light brown sugar

200ml organic apple cider
    vinegar

1 star anise

1 cinnamon stick

1 tsp ground cumin

1 red chilli, halved and
    deseeded

100g sultanas

Sharing board

4 thick slices Yorkshire ham

2 red apples, sliced

150g Wensleydale cheese

3 sticks celery, cut into batons

150g grapes

1 sourdough loaf

Yorkshire butter

METHOD

1. To make the chutney, put all the ingredients into a large, heavy-based pan and stir well to combine. Bring slowly to the boil, stirring occasionally, then reduce the heat and simmer for about 2 hours, until the chutney has thickened. Remove the chilli, cinnamon and star anise, and discard.

2. Carefully spoon the chutney into warm, sterilized jars (see tip below) and seal, label, then store for at least 2 weeks to mature, before eating.

3. When ready to serve the sharing board, arrange all of the ingredients on a large board, along with the chutney.

Tip

To sterilize jars: wash out jars with hot, soapy water, then rinse. Half-fill each jar with water, place in the microwave in batches of 4–6 (depending on size of microwave) and bring to the boil on full power. Remove carefully, pour out the hot water and place the jars upside down on a clean, dry tea towel to drain, ready for filling.

Or, heat the oven to 150°C/130°C fan/gas 2. Wash out jars with hot, soapy water, then rinse. Place the jars upright on an oven tray and heat in the oven for 20 minutes. They will then be ready for filling.

Fill the hot jars carefully (using a jam funnel if you have one) to within 5mm of the top of the jar. Make sure the rim of the jar is clean before screwing the lid on tightly. Label when cool.

Raspberry Jelly

Prep time 5 minutes / Cooking time 15 minutes / Makes 1 x 400g jar

INGREDIENTS

400g raspberries, washed
    and dried

300g preserving sugar

2 tsp lemon juice

METHOD

1. Put the raspberries into a heavy-based pan and roughly mash them with a potato masher. Sprinkle over the sugar, add the lemon juice and give everything a good stir.

2. Gently heat the pan on a low heat for about 10 minutes until the sugar dissolves, stirring occasionally to ensure that the sugar is fully dissolved.

3. Next, place a fine metal sieve over a bowl and pour the hot mixture through the sieve; you will need to press down firmly with a metal spoon to get all of the raspberry pulp through the sieve, leaving the seeds behind.

4. Return the raspberry pulp to a clean pan and bring to a rolling boil for 5 minutes. To test if the jelly is ready, place a spoonful of jelly onto a cold saucer, then run your finger through after a minute. If the jelly has set the surface should wrinkle and be fairly firm. If the jelly is not quite set, then boil it for another minute.

5. Put the raspberry jelly into sterilized jars (see tip below), seal with a lid immediately and allow to cool before storing in a cool dark place. Once opened, keep refrigerated.

Tip

To sterilize jars: wash out jars with hot, soapy water, then rinse. Half-fill each jar with water, place in the microwave in batches of 4–6 (depending on size of microwave) and bring to the boil on full power. Remove carefully, pour out the hot water and place the jars upside down on a clean, dry tea towel to drain, ready for filling.

Or, heat the oven to 150°C/130°C fan/gas 2. Wash out jars with hot, soapy water, then rinse. Place the jars upright on an oven tray and heat in the oven for 20 minutes. They will then be ready for filling.

Fill the hot jars carefully (using a jam funnel if you have one) to within 5mm of the top of the jar. Make sure the rim of the jar is clean before screwing the lid on tightly. Label when cool.

Raspberry and Almond Scones

My favourite scones are moist enough in their own right that just a spreading of butter does the trick. Although the jelly is of course a lovely addition. In my opinion, my scones are so good I don't let anyone else in the house make them!

Prep time 10 minutes / Cooking time 15 minutes / Makes 8

INGREDIENTS

300g self-raising flour, plus
   extra for dusting
1 tsp baking powder
85g butter, softened and cut
   into cubes, plus extra for
   greasing
85ml milk
1 egg, beaten
50g caster sugar
100g fresh raspberries
30g flaked almonds

To serve

whipped or clotted cream
homemade raspberry jelly
   (see page 173)

METHOD

1. Preheat the oven to 220°C/200°C fan/gas 7.

2. Sift the flour and baking powder into a mixing bowl, then add the butter and rub in with your fingertips until the mixture resembles fine breadcrumbs.

3. Measure the milk into a jug, add the egg and whisk with a fork to combine.

4. Stir the caster sugar, raspberries and two-thirds of the almonds into the flour, using a round-bladed knife, then add enough of the milk mixture to form a soft dough.

5. On a lightly floured surface, shape the dough into an 18cm round. Cut the scone into eight wedges and place on a lightly greased baking tray. Brush each scone with any of the remaining liquid then sprinkle over the remaining flaked almonds.

6. Bake for 13–15 minutes, until well risen and golden. Transfer to a wire rack to cool slightly, then serve warm with cream and raspberry jelly.

Tip
If you prefer, substitute the almonds with white chocolate chips.

September

By now the lambs have morphed from spin-dly, skittish, dependent little beings into an emboldened, inquisitive, adolescent phase. During their months at the moor they have filled out, grown stronger and more confident, venturing further from their mothers each day. It is time for them to be 'spained', or weaned, and so we gather them up and bring them down to the meadows. These are now lush and green, and full of goodness, having enjoyed a good month rejuvenating themselves after haymaking time. It truly has become a case of the grass being greener on the other side of the fence for the young stock, who once through the gate graze the fog to their hearts' content. They are losing their nutritious milk supply, but the bright green regrowth compen-sates for this perfectly.

Meanwhile the yows are left in peace to recover their strength. Pregnancy and lactation has inev-itably taken its toll, and they need now to regain their condition. It was a bonus that breastfeeding my children, coupled with the amount of physical exercise I was doing around the farm, kept me slim, but for the sheep that is not the goal! They need a layer of fat on their backs – we want them to be hearty, ready to start the motherhood journey again, with tupping time just two months away.

There's another important reason to wean the lambs at this stage. The tup lambs are now suffi-ciently developed that they have male hormones coursing through their veins, and the last thing we want is any inbreeding. So separating the lambs from their mothers and sisters in September is a good way to ensure that they don't get too frisky with each other.

The highlight of September is Muker Show, our local agricultural show. It falls on the first Wednes-day of the month, without fail, and feels like a real holiday, with all the family taking the day off to enjoy the activities. There is everything you would expect from a country show, from sheepdog trials and stick carving to vintage tractor and cake-baking competitions, and all set against the spectacularly beautiful backdrop of Muker Meadows.

For one day only the children run amok on the showfield. If I want to find the younger ones, my best bet is the bouncy castle. Clive slips the man overseeing it a few pounds to let the children have unlimited access for the day.

The sheep show is taken extremely seriously, especially by Swaledale enthusiasts, who can't be moved from the sheep pens for the duration. It is an opportunity for farmers and breeders to show off the cream of their crop and the animals that

they will be selling later in the autumn, whetting the appetite of would-be buyers.

By mid-afternoon, attention turns to the Fell Race, which starts in the middle of the show-field. From there the course follows a footpath, diverts through a river, and heads up the fell that dominates the skyline, an upwards struggle that steepens so significantly that, as you reach the summit, it is necessary to crawl. Once at the top the race is nowhere near over, as the fell runners then begin the descent back down to the start point, with the best ever time somewhere around the thirteen-minute mark. It is practically obligatory for all the young harriers in the district to participate in this famously difficult race, although the winner will inevitably be an older gentleman, whippet-thin with sinewy legs, cotton shorts, and basic trainers. Experienced fell runners are certainly a unique breed!

Every year I would help the children change into their running gear, fill in the consents, pin their race cards to their vests and then cheer them on in their pursuit of glory. Miles is a particularly good runner, and has won the junior section twice in a row, and is now holding out to complete the hat trick. Then in 2019 I decided that as I'd been pregnant and/or breastfeeding for the last twenty years, it was time to start a tougher fitness regime with a view to competing in the ultimate race. It coincided with the theft of our two quad bikes, when I resolved I would walk or ride a horse instead of replacing the quads, and to really get things moving, at the same time I took up running. So naturally my mind turned to making my competitive debut at Muker Race . . .

I knew I wanted to have a go, and that I didn't have any hope of actually winning in any category. There was one prize for 'first competitor home, living in district', but unless they extended that to 'first mum-of-nine home, living in district' I still stood no chance of being placed. This was more of a personal challenge; a need not to win, but just to run it, and come back alive, to prove to myself that I might be in my mid-forties by now, but I wasn't past it. But as the show day approached, I started imagining the worst and became fearful of some disaster befalling me. What if I was sick halfway round the course, had to limp back or, worse still, just couldn't physically do it?

The evening before the show, as is tradition, we went down to the ground with the tractors and entries for the produce competitions and to help get everything set up. Despite the drizzle and low-lying mist, the field was a hive of activity, with marquees already erected and filling with submissions for the next day. By this point I had come up with a cunning plan. The race route would have been laid out, the flags to guide the runners in place, so there was no reason why I couldn't run it alone that very evening. That way there would be no spectacle, I couldn't be humiliated, and I couldn't lose as I was only competing against myself.

'Right, Clive. I've got my gear, so while you find out from Ernest [the show organizer] where he wants the vintage tractor, I'm gonna do this run.'

'Yer a daft bat, thee!' he said, shaking his head.

'Mebbe, but I'm doing it,' I retorted.

So after surreptitiously making my way past the exhibitors and organizers, I set off in my shorts and vest top, with well-worn trainers on my feet.

The first stumbling block was the river. Normally the runners splash through it up to their knees but there had been heavy rain recently and, grimacing at the cold, I went in, wading through the waist-deep water. Not a great start, but I was going to get wet anyway since it was drizzling and I knew I would soon be overheating, so I pushed on.

In no time at all I was into the mist, gasping for breath, my hair sticking to my face through a mixture of snot, rain and sweat. I focused on the route ahead. It was an incline like you wouldn't believe and I had to give myself a serious talking to, to keep pushing on. As I scrambled up the last ridge to the top, dodging rabbit holes and clumps of crumbling earth, I felt like I had really achieved something, but there was no time to waste as the most disconcerting part of the race – the descent – lay ahead. To run downhill requires bravery and complete confidence in your stride and, as a novice, I was nervous of taking a heavy tumble. I slalomed back and forth along the course, not watching the clock, just focusing on my reddened, painful legs and heavy breathing. I came back through the low cloud to see the river below, but the relief setting in was abruptly halted by the sight of Clive teetering on rocks in the middle of the river. He was attempting the crossing in his wellies, with his arms outstretched. Before he caught sight of me he took a large step – a misjudged one at that – which saw him lose his footing and fall right in. For a moment he disappeared from sight, completely submerged, and then surfaced; coat billowing, he spluttered and readjusted his peaked cap, which had miraculously stayed in place. By this time I had reached the river myself and was less bothered about getting wet than I had been on my outward journey, so I waded back across confidently, following a scowling Clive, who had apparently begun to fret about how long I had been gone and decided that I must have fallen and broken my leg (or succumbed to altitude sickness . . . !).

To say he was raging would be an understatement. He stalked back towards the marquees, sopping wet and squelching all the way. I followed meekly a few steps behind. He paused for a moment to take off a welly and tip out the water.

'Why can't yer just be bloody normal?!' he suddenly burst out, as everyone on the showfield went about their business, scurrying around with their entries, trying their utmost to pretend they weren't nebbin.

I didn't know whether to feel elated from the run, dejected by Clive's comments, or laugh at how funny the situation was. In the end I opted for keeping quiet. We sat in silence on the journey home, the car heater on full blast. By the time we had driven a couple of miles the windows had steamed up with condensation due to our sodden state.

'What 'appened?' enquired Miles as Clive tried to remove his socks in the doorway, leaving two wet footprints. Standing behind my husband, I gesticulated to Miles to shut up. The incident was never spoken of again.

It gave me enough confidence to decide that I could run the course for real the following year without embarrassing myself in front of the gathered throng. Of course, the pandemic scuppered this plan. It was inevitable that the show would have to be cancelled and we were obviously very disappointed. In a bid to keep up the enthusiasm and momentum we decided to create our own Fell Race at Ravenseat instead. It was a dreadfully wet and windy day, but everyone from Sidney's age upwards took part in the challenge to run up one side of the Hoods Bottom Beck valley to a flag that Reuben and Raven had planted at the top of the Graining Scar waterfall, and back down the other side. The beauty of our race was that we

Before school begins we would hope to have enjoyed rich roadside pickings of hedgerow fruits, the most loved, of course, being brambles.

could all choose our own line, whether that be to climb out onto the tops early on and run the watershed, or to conserve energy and climb the steeper, more treacherous screes later in the race. The boys outran the girls; I limped back not quite last, tired but thoroughly elated.

Before school begins we would hope to have enjoyed rich roadside pickings of hedgerow fruits, the most loved, of course, being brambles. We have our favourite places to visit but the bounty varies; a proliferation one year, and meagre offerings the next. Nevertheless, the children's enthusiasm for being stung and prickled never seems to wane as

they appear with baskets overflowing with dark red and purple berries, and the odd caterpillar. Scraped arms and stained mouths, grins complete with seedy teeth, prove that not all of the harvest goes into the basket.

We have a small area of brambles that grow beside a lime kiln at Anty Johns and a limited amount at The Firs too, in a copse beside the beck, but generally we have to make a little more effort to go blackberrying, getting in the car to travel towards the climatically kinder, always more fruitful Eden Valley. There is so much out there to be had for free, in exchange for a little time and effort.

Some of the fruit goes straight into crumbles, pies or jellies. All hedgerow fruit, even the jarringly bitter rowan, is tolerable with the addition of copious amounts of sugar. Rosehip jelly is a lovely

accompaniment to bread and cheese, while rowan jelly is perfect with gamey meat.

I make a concerted effort to put some fruit aside to go in the freezer, ready to be used in the winter months that lie ahead. In a farmhouse kitchen the aim is to make the most of the summer glut through preserves and freezing, but it can be tempting to get excited and use it all there and then, leaving nothing to brighten the diet in January. I sometimes think our enjoyment of this produce is less about the taste and more about the memories evoked of harvesting on a beautiful autumn day, and the fun of baking.

The one thing we aren't fortunate enough to have is an orchard. The acidic soil and harsh weather just does not suit larger fruiting or parkland trees but, once again, there are many places nearby where it is easy to pick up surplus fruit. Stored in the cool dairy, they stay good for ages.

A recent haul made us a lovely apple dappy – spirals of dough, like a Chelsea bun but with an apple filling – finished off with a custard made from milk, courtesy of Buttercup. After a glut of apples one year, we made a particularly unctuous concoction I can only liken to Jersey Black butter. Spiced with liquorice and cinnamon, the apples were caramelized and simmered for a very long time; it was delicious on bread, or even used in a tart.

Baking is certainly not undertaken by me alone. The children like to be involved too, especially with any produce they have foraged for themselves, and I actively encourage and challenge them to make meals from scratch. The dairy is well stocked with the staples – big sacks of sugar, dried fruit and flour – from which cakes, buns, pastries and pies can be created. Only the chocolate is hidden from their keen hands. I have a very heavy, black, iron goose pot with a lid so weighty that they can hardly lift it off, so the chocolate lives in there.

I will help them with their baking if they ask, but I would rather leave them alone, as it is better that they work through the recipe themselves. When all goes well, the hunger-inducing smells of flapjacks, bread or tiffin start to waft temptingly around the farmhouse. And if it doesn't go quite as smoothly, well, they will learn from their mistakes. Raven made a cake once but used the wrong kind of flour – an easy mistake to make – and of course it didn't rise. Edith made Bramley apple muffins, but didn't preheat the oven, so the mixture had well and truly stuck to the cases before we could put them in to bake. Nothing ventured, nothing gained; there may be mess in the kitchen, and occasionally a leaden loaf or two, but this is the start of the children's culinary journey and there is no substitute for hands-on learning. In time they improve and we reap the rewards.

All hedgerow fruit, even the jarringly bitter rowan, is tolerable with the addition of copious amounts of sugar.

Moroccan Lamb Tagine with Jewelled Couscous

This is a real staple in the Owen household. I love combining fruit with meat, both for the flavour and the colour. You don't need a tagine to make this meal – I just use a large casserole dish. Once you have done the preparation, you can happily leave it cooking for an hour or so and get on with other things.

Prep time 15 minutes / Cooking time 1 hour 15 minutes / Serves 4

INGREDIENTS

Tagine

2 tbsp olive oil

600g lamb shoulder, diced

1 large onion, sliced

2 cloves garlic, crushed

1 tbsp ground cumin

2 tsp ground coriander

1 tbsp hot smoked paprika

40g tomato puree

400g can cherry tomatoes

1 cinnamon stick

50g dried apricots, chopped

2 large pieces orange peel

285ml lamb stock

1 small bunch coriander, chopped

Jewelled couscous

200g couscous

200ml hot vegetable stock

1 red onion, chopped

100g pomegranate seeds

1 small bunch mint, leaves roughly chopped

1 small bunch coriander, chopped

3 spring onions

20g almonds, toasted

salt and black pepper

METHOD

1. Preheat the oven to 180°C/160°C fan/ gas 4. Heat 1 tablespoon of the oil in a large heatproof casserole dish, then add the lamb and sear and brown on all sides for about 5 minutes. Remove the lamb from the casserole dish with a slotted spoon and set aside.

2. Return the casserole dish to the heat and add the remaining oil. When hot, add the onion and fry gently for approximately 8 minutes until golden. Add the garlic and fry for a further minute.

3. Sprinkle over the ground spices and stir in the tomato puree then fry for a further minute. Return the meat and any juices to the pan with the remaining tagine ingredients (except the fresh coriander), bring to the boil then cover with a lid and transfer to the middle shelf of the oven. Cook for 1 hour, or until the meat is tender.

4. Meanwhile, while the tagine cooks, put the couscous into a bowl and pour over the hot vegetable stock. Cover the bowl tightly with cling film and set aside for 10 minutes until the couscous has absorbed the liquid. Season the couscous, then stir through the remaining ingredients, saving a few almonds and herbs to garnish.

5. To serve the tagine, remove the cinnamon stick and check the seasoning, then sprinkle over the chopped coriander. Serve in warm bowls with the jewelled couscous.

Tip
This recipe can also be cooked in a tagine over a barbecue.

Hedgerow Nutty Crumble

Prep time 10 minutes / Cooking time 40 minutes / Serves 4

INGREDIENTS

Fruit filling

40g butter

40g golden caster sugar

2 Braeburn apples, peeled
   and sliced

200g Bramley apples, peeled
   and sliced

300g foraged berries or
   any mixed berries, such
   as blackberries and wild
   raspberries

Crumble topping

125g plain flour

85g unsalted butter, diced

85g demerara sugar

50g rolled oats

30g chopped toasted
   hazelnuts

custard or ice cream to serve

METHOD

1. Preheat the oven to 180°C/160°C fan/gas 4.

2. For the fruit filling, melt the butter in a pan, add the sugar and apples and gently simmer for 5 minutes. Add the berries and simmer for 2 minutes, then transfer to a 23cm round ovenproof dish.

3. Meanwhile, prepare the crumble topping. Put the flour into a mixing bowl and rub in the butter with your fingertips, until the mixture resembles fine breadcrumbs. Stir in the sugar, oats and hazelnuts.

4. Sprinkle the crumble topping over the fruit, then place the dish on the middle shelf of the oven. Bake for 30–35 minutes, until the crumb is golden and fruit bubbling.

5. Allow to cool slightly before serving with custard or ice cream.

Tip
Try using other seasonal fruits, such as damsons or gooseberries.

October

By October we can no longer rely on being woken by the sunlight creeping through our curtainless windows, and our days now begin and end in the dark. The shortening daylight hours, and definite nip in the air, are a sure sign that the long winter months are hovering nearby, ready to swoop in.

Our time is spent on the 'Harvest of the Hills', which began in September and runs all through October. 'What do you harvest though? You don't grow any crops!' is the common question. Indeed, but we do grow and harvest sheep in a way. I'm not talking about harvesting them for the freezer, as it wouldn't make economic sense for us to be rearing animals for meat here. We have a short grazing season and a long winter, so fattening animals to go into the food chain would be hard work. Our main focus is rearing animals to then sell on to other farmers as breeding stock, called seed stock in America, which seems quite apt. In a way we get the nice job, the early part, the creation of life.

Everything we breed up here – whether that is children or sheep! – is exceptionally hardy. You have to be, to survive the conditions. So as the sheep get older and move down to kinder climates in the lowland farms, they really thrive. There is often an assumption when spotting lorries full of animals on the motorway that they are on a one-way trip, but that is not always the case. They may well be off to the flatter, sunnier counties of Devon or Cornwall, for an easier life.

The sheep we're selling are all brought down from the moors, sorted into categories, cleaned, stray hairs plucked, and generally beautified to a greater degree than a bride on her wedding day, then sold at auction, yows first, and then tups. A buyer will look at their teeth and testicles, and can be spotted squeezing the top of the tail, to make sure it is thick, for that extra protection against the elements. It can be quite a stressful or exciting time, depending on what kind of year we have had with the sheep.

I have always felt satisfied with the life that they lead both with us and when they head on to the lowland farms, apart from one group, whose fate used to trouble me. Each year the majority of the lambs are either sold to be bred elsewhere or we decide to keep them to breed here. The only ones that don't fit into either category are the male lambs who don't meet the breed standards and are then sold for meat. Before last summer, I would take them to auction and sell them there, but I always felt a niggling concern that I didn't know their exact fate once they left the show ring.

There was always a chance they could go for live export, and the truth is that troubled me. To know the next step in their journey felt important, so I decided the only way to be sure that it was one I would find acceptable was to take control of it.

I spoke to Martin, a butcher friend at the local abattoir, about setting something up. Let's not pretend that anybody likes abattoirs. No one wants to think about what goes on there, including me. But if you are going to eat meat, you need to accept their existence and rather than blocking it out, take an interest in the process and ensure it is the best it can be.

I am fortunate to have this abattoir on my doorstep. It is a forty-minute drive from the farm, so there is no risk of stressful long journeys, and loading on and off at docks. On arrival there isn't a long wait in a production line, and I know how everything is carried out. Having that knowledge sits better with me.

So now, on a night, Martin tells me how many lambs he wants. In the morning I go in and feel their backs, checking for a certain weight and specification. There is a lot of thought that goes into choosing the lambs, and there is a grading system, whereby you want some fat on them, to keep the meat tasty and succulent, but not too much fat. People's tastes have changed in recent years towards leaner meat, which our lambs are perfect for. A life led out on the hills means they have been moving and active, grazing and working to find a morsel of grass rather than ambling over to a feed hopper.

I drive the lambs over to Martin, and by the evening I have the weights. Hill breeds will never yield a great weight of meat, the kill-out ratio being, percentage-wise, considerably less than a butchers'-bred lamb, but what you do get is exceptionally tasty and significantly healthier.

It is interesting getting the grades back as it helps me learn which tups to choose and what people want. People ask if I know everything there is to know about sheep as a shepherdess, but I am always learning, and adapting, and always will be. This is just one of those ways.

People can now order a lamb from Ravenseat that is cut to their specification and delivered in a box to their home. The nearer the meat is to the bone, the sweeter it is. We weren't sure how this service would take off when we launched it in the summer, but by the autumn when the lambs were maturing, they were selling like hotcakes. It felt like a satisfactory solution, and more in line with our ethos. The idea of keeping it all local is closer to how things used to be. I realize not everyone can afford to buy lamb this way, either due to cost, freezer space, or an understanding of how to cook it, but the option is there.

Any other animal that we keep for our own freezer also goes through the same abattoir, as we never butcher them ourselves. We eat our own lamb, happy in the knowledge they have had a good life out on the moors, but whether we keep any of the bullocks depends very much on trade – if it is good, we always sell.

If the weather has been damp and humid, October can see a sudden flurry of white umbrellas popping up across the fields, as mushrooms decide it is their time to shine. There is an ongoing family joke that I can spot one at a hundred paces, hiding in amongst the long grasses.

The horse mushroom, with its fleshy white cap and brown gills, and an unexpected aroma of

aniseed, is the most common type found at Raven-seat. They don't grow on the moor, but prefer the permanent pastures, getting their name from their love of manure – something we have no shortage of here. They like to make their home near a wall or fence, as though they are trying to find a quiet spot, and have a helpful habit of coming back in the same place from year to year, so if I know there was a good supply in the corner of one particular field, chances are I'd do well to revisit the same spot the next year.

I chop the mushrooms, fry them off with garlic, and enjoy the comforting, earthy flavours, or make a really thick, hearty soup, great to dunk crusty bread into. Raven makes a lovely risotto, packed with mushrooms and Parmesan.

Orange chanterelle sometimes make an appearance, their meaty texture and almost fruity flavour making a great side dish.

For some unknown reason puffball mushrooms like to make their home amongst the nettlebeds. Perhaps they hope the stinging leaves will offer them some protection, but having perfected their nettle-pulling skills earlier in the year, the children are not deterred. They will appear through the kitchen door carrying the large, spongy, white balls as an offering for dinner, Violet even once arriving with one as large as her head.

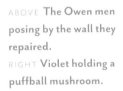

ABOVE **The Owen men posing by the wall they repaired.**
RIGHT **Violet holding a puffball mushroom.**

There is no great skill to foraging, just a thorough knowledge of exactly what you are looking for. I know my patch well, but the woodlands would be another matter, with a completely different set of fungi. If I spot a mushroom and I am not 100 per cent sure of its type, I will always leave it be. It is wise to always err on the side of caution, as of course there are some poisonous types in the UK.

I don't like to waste food, so I do occasionally dry mushrooms when I have a bounty, but my one attempt at freezing them taught me that was a mistake. More often than not, we just embrace the season, and eat them with everything, overdoing it, until the mushroom's time is over, and they are dying out. Some years we have such a glut of mushrooms that the children turn entrepreneurial and lay them out on a table, asking anyone passing through to help themselves in exchange for a donation.

I had a couple of visitors staying in the Shepherds Hut and only found out they were vegan

when they arrived. Unprepared, and lacking in options for them, they were presented with grilled mushrooms on toast every morning. Thankfully it was mushroom season, and they were more than happy to start the day with those fabulous, filling flavours.

Then of course October ends with Halloween. Unsurprisingly, given the rich history of the moors, there are plenty of ghost stories and folklore about the area. I am unsure if I believe in ghosts or not, but I did have an experience at The Firs that to this day I can't make sense of. The Firs is the farmhouse that we own and let to holidaymakers, a couple of miles away from Ravenseat in Upper Swaledale. I was there on my own one dinnertime when we had no guests, and I was after somewhere quiet to join in an online talk as part of a book festival. It was daylight, and The Firs is always a place where I feel relaxed and happy to be alone, so the supernatural was the last thing on my mind. It was a drab day, and I lit the fire, sorted my laptop and notes and settled back, but as I was about to log in, I realized the internet wasn't working. Getting Wi-Fi to The Firs has been a nightmare, and we've spent a fortune getting all sorts of cables, and signals bouncing off who-knows-what to achieve it, so I felt slightly irritated. The only thing for it was to head back to Ravenseat and do it from there – far from ideal with all the children running about. Book fair attendees might be amused once by a child running past naked, or a dog jumping into my lap for attention, but there was definitely potential for that to dominate the entire talk.

I sat back and sighed, taking a moment to compose myself before the drive back. A grandfather clock began striking, and I listened to the sound, counting along with it, until it suddenly registered . . . there wasn't a grandfather clock in

I got a real case of the chills, and packed up quickly, to head back to the safety of the farmhouse.

The Firs. I looked round, trying to find the source, but there were no other clocks, and the TV and radio weren't on. It had definitely been the distinctive chimes of a grandfather clock.

I got a real case of the chills, and packed up quickly, to head back to the safety of the farmhouse. It might not have been an obvious supernatural sighting of the headless-person-carrying-a-candle type, but I could not find an explanation.

Raven avoided The Firs for a while too, for a different reason. A friend had told her that years ago someone had died in the upstairs bedroom, been laid out up there and then put in a coffin. When it came to moving the body out of the house, they couldn't get the coffin down the stairs. There is an old stone staircase with uncovered, worn stairs, and a sharp bend partway down, so I can well believe someone failed to see the practical issue ahead!

I enjoy hearing local tales that are passed down, even knowing there's embellishment added each time they are told. But out on the moorland in the dark, looking for a lost walker, it can become more of a horror story. A gamekeeper shot by poachers on the ground where you are standing, a hundred years ago? Interesting to hear about it in the sunshine, something you want to forget when out there at night.

Wild Mushroom Soup with Wood Sorrel and Hazelnut Pesto

Think of mushroom soup, and you imagine the grey mush of childhood. Thankfully this version is full of texture; just be sure not to overblend. The toppings add really complementary flavours and brighten the dish.

Prep time 10 minutes / Cooking time 35 minutes / Serves 4

INGREDIENTS

2 tbsp Yorkshire rapeseed oil, plus extra for garnish

1 medium onion, finely chopped

1 large clove garlic, crushed or grated

450g wild mushrooms, such as yellow chanterelle, roughly chopped

50ml Madeira or sweet sherry

850ml hot vegetable stock

3 sprigs fresh thyme

200ml double cream

1 tsp mushroom powder

salt and black pepper

Pesto

25g chopped toasted hazelnuts

1 small bunch wood sorrel

2 tbsp olive oil

10g grated Parmesan

Garnish

handful of wild mushrooms

METHOD

1. Heat the oil in a large saucepan, add the onion and garlic and fry gently for 5 minutes, until softened but not coloured.

2. Next, add the mushrooms and fry on a medium heat for 8 minutes, until they are golden.

3. Pour the Madeira over the mushrooms and cook on a high heat for about a minute until the alcohol evaporates.

4. Lower the heat and pour in the hot stock, then add the sprigs of thyme. Bring to the boil then cover and simmer gently for 20 minutes.

5. Remove the pan from the heat, blitz with a hand blender or in a liquidizer until the soup is smooth. Return the soup to a clean pan, stir in the cream and sprinkle over the mushroom powder, then season the soup and gently reheat over a low heat.

6. Meanwhile, place the pesto ingredients into a small food processor and blitz for a few seconds until combined. Add extra oil if the pesto is a little thick and season to taste.

7. For the mushroom garnish, drizzle some oil in a small frying pan and gently fry the remaining mushrooms for a couple of minutes until golden.

8. To serve, ladle the soup into warm bowls, scatter over the whole fried mushrooms, and top with a spoonful of pesto.

Tip
Try serving with some homemade spelt bread.

Spelt Bread Wedges with Red Onion and Rosemary

Prep time 10 minutes / Cooking time 35 minutes / Makes 4

INGREDIENTS

1 medium red onion, peeled
and finely chopped

2 sprigs rosemary, leaves
roughly chopped

1 tbsp Yorkshire rapeseed oil

170g spelt flour, plus extra for
dusting

2 tsp baking powder

2 tsp wholegrain mustard

180g potatoes, peeled and
coarsely grated

1 medium egg, beaten

2 tbsp milk, plus a little extra
to glaze

salt

METHOD

1. Preheat the oven to 200°C/180°C fan/gas 6.

2. Put the red onion and rosemary on a small baking tray, then drizzle over the oil. Place on the top shelf of the oven and cook for 10–12 minutes until the onion has softened, then set aside to cool.

3. Heat a baking tray in the oven. Place the spelt flour, baking powder, mustard, grated potato and onion into a bowl and mix well. Season with salt.

4. Make a well in the centre of the flour mixture, pour in the egg and milk, and mix to combine.

5. Turn the dough out onto a lightly floured surface, gently knead together and shape into a 15cm round. Cut into four wedges then place on the hot baking tray. Brush the tops with milk and bake for 25 minutes until well risen and golden.

Tip
Serve warm with butter. The spelt bread wedges are best eaten on the day they are made.

Romany Lamb Stew

I was taught how to make this stew by an elderly Romany woman the first time I visited Appleby Horse Fair. It is full of flavour and I love the addition of the barley to thicken it. I've adapted it over the years, and feel free to do the same yourself.

Prep time 15 minutes / Cooking time 1 hour / Serves 4

INGREDIENTS

2 tbsp Yorkshire rapeseed oil

450g lamb shoulder, diced

1 medium onion, diced

2 cloves garlic, crushed

1 stick celery, chopped

1 medium leek, sliced

450g potatoes, peeled and diced

1 small swede, peeled and diced
　　(about 300g)

1 medium carrot, peeled and
　　sliced

1 medium parsnip, peeled and
　　sliced

3 tbsp plain flour

1 tbsp smoked paprika

100g barley, rinsed and drained

800ml lamb stock

2 sprigs rosemary

1 small bunch thyme

400g can haricot beans, drained
　　and rinsed

100g frozen peas, defrosted

1 small bunch fresh parsley,
　　chopped

salt and black pepper

METHOD

1. Heat half the oil in a large heatproof casserole pan, then add the lamb and cook for 5 minutes to brown and seal the meat. Remove the meat from the pan to a bowl with a slotted spoon.

2. Add the remaining oil to the pan and gently fry the onion for 5 minutes until softened then add the remaining vegetables and continue to cook for 8–10 minutes, until softened.

3. Next, return the lamb to the pan along with any meat juices, sprinkle over the flour and paprika, and cook for a minute over a medium heat.

4. Add the barley and pour over the lamb stock. Stir through the rosemary and thyme sprigs, bring to the boil, then cover and simmer for 35–40 minutes until the lamb is tender.

5. Stir in the haricot beans and peas, and heat gently for 5 minutes.

6. Finally, season to taste, then sprinkle over the parsley before serving.

Tip
This is a one-pot meal but you could serve it with some crusty bread to soak up the delicious juices.

November

We are by now descending headlong into what we call the 'back end' of the year. The days are short, even with the end of British Summertime and the clocks being put back an hour to give us that all-important extra light in the morning. Soon the children will leave for school under the cover of darkness and return to the farm when it is cloaked in an inky blackness but, for now, we are still able to pack all our daily tasks in before nightfall.

With the autumn sales over it is full steam ahead with sheep breeding, which in the hills is late; Bonfire Night is early enough to even consider turning any tups out. A lamb conceived now will be April-born, which, with the late arrival of spring in these parts, is sensible. Nothing will be gained by trying to beat the season's clock. A lamb born any earlier will require so much more in terms of sustenance, care and shelter from the brutal elements that there is no benefit to be had.

From tupping time onwards, the sheep are our main priority. The hill shepherd's calendar is split almost perfectly in half, with six months of absolute devotion to the flock, which are foddered and moved on a daily basis, and then six months of our charges being free to wander the moors without hindrance, gathered up only on specific occasions for clipping or speaning (weaning).

Tupping time sees every single sheep brought down from the moors and sorted up into smaller flocks to be paired with a tup that we think will complement them – the idea being that we improve our flock year on year through breeding better replacements. Each small flock of sheep must be visited every day and gathered up to the tup. This is to make sure that all of the yows in season that day are tupped (mated). If this opportunity is missed then it will be another seventeen days before the yow cycles again, and thus lambing time is delayed. The task of going from field to field to apply ruddle to the tups can seem like a thankless endeavour. After all, it has been said that 'It ain't ruddle that gits thi a lamb'! This criticism is usually cited after an overly liberal application of the oily paint that transfers from the tup's brisket onto the yow's rump. It is easily done, surrounded as I am by sheep, holding in one hand a scoop of feed that the tup is eating, and in the other a wooden paddle daubed with dripping, brightly coloured ruddle, the pot of which is nipped between my knees. All the while the sheepdog is circling, holding the flock tightly around me. There is a lot to watch and undoubtedly it is easier

with an extra pair of hands, even if they end up being stained to match the bright colour of the ruddle. The children are keen to help and aren't so bothered by the elements. There is no such thing as bad weather, only wrong clothes, though it feels as if a long haul is ahead of us as we don the waterproof attire that we will wear almost constantly until at least lambing time.

If the weather treats us kindly then it is delightfully satisfying to be outside working with your sheepdog under the heavy iron skies, the purple-hued moorland vista bathed in a soft autumnal light. If the heavens open, the land becomes wet and each gateway is a sea of mud, and it can be a real slog to do the rounds, visiting some ten separate flocks of sheep. The flock will be in good

With the autumn sales over it is full steam ahead with sheep breeding, which in the hills is late; Bonfire Night is early enough to even consider turning any tups out.

will feel dispirited if there is no let-up in the rain. The feeling is mutual – I might be wrapped up in my waterproofs and keeping dry, but it's difficult not to be despondent as I traipse through mud with incessant rain driving at me.

The sooner the yows are in lamb the quicker they can return to their heafs at the moor. After a month of being clear of grazing stock whilst the yows were in-bye, the moor will have freshened and, with its varying terrain, it offers the sheep naturally sheltered areas. I'll always attempt to find a place that is dry underfoot when foddering the sheep, for the land soon becomes soiled when there is a hungry flock vying to be the first to get a mouthful of the summertime bounty. I'll try to gauge the wind direction, too. It is difficult to fodder the sheep when the weather is clashy and the wind picks up the hay and blows it around in a maelstrom of wispy, bitty stalks that get everywhere. It is such a waste – but even worse is the extreme irritation and damage hay seeds can cause to the sheep's eyes. When compounded

bodily condition, a month after being speaned from their lambs and now fully fleeced. Their woolly coats are like wearing a jacket and an over-coat in one, with a fleece of tight-knit finer wool next to the skin, then an outer layer of coarser, kempy wool. Even on the worst day you can part the fleece and they will still be dry next to their skin. They will not feel the cold, but I am sure they

by rain, infection will soon take hold. The first sign of a problem is swollen eyelids and redness around their eyes, which then start to look sore and milky. Within days the afflicted sheep will lose their peripheral vision and, if not treated quickly, complete blindness will follow. Their peculiar stance – neck outstretched, lugs pricked and head tilted up to listen – and the way they develop an awkward high-stepping gait to avoid any potential pitfalls are the most obvious indicators of the onset of the blind illness. If left, the eye can ulcerate and they will lose it. It is a depressing state of affairs, but if dealt with quickly and efficiently the problem will only be temporary. It is just a matter of catching them first! Sheep afflicted with blind illness are blessed with super-sensitive hearing and a new-found sixth sense that tells them exactly where and when danger is approaching.

I was bothered with what I can only assume was the same ailment: sore, dry eyes and puffy red eyelids. A trip to the pharmacist got me a couple of tubes of eye cream that treated me and, when I was cured, a few of my flock too.

Wetlands dominate our moors, covered with cotton grasses, sedges and many varieties of sphagnum moss. The moss creates a living carpet of lurid green, in some places so dense as to form mounded hummocks of a spongy mass, giving a clear indication that the ground is waterlogged. This bog-loving moss can soak up to eight times its own weight in water. In the areas where we have blocked up the drainage grips, this is exactly what we want to see. But in terms of navigating our land, the general rule of thumb is that the brighter the green of the moss, the deeper the bog beneath it, so avoid those areas unless you want to experience that sinking feeling.

To discern the best route over the moorland, you should heed the old saying: 'The seavy swang ne'er lost horse nor man.' Seaves are a reference to the rushes, the swang being the bog. The roots of the rushes give a foothold, so even though they grow in wet areas they will take your weight and prevent you sinking entirely.

Sphagnum moss has great antiseptic qualities, and was used during the First World War to treat wounds. It was also (and this idea I applaud) used by Native Americans who lined their children's cribs and carriers with it as a natural nappy! If only I'd known. For the same reason it also apparently makes a great deodorant. Mind you, the same is said of wool but there are definitely some stinky sheep out there!

We've always looked for ways to be even more self-sufficient, and as winter approaches the logistical challenges of our geographical location are even more apparent. The road over the county boundary between Cumbria and Yorkshire becomes a no-man's land, and the road between us and Kirkby Stephen cuts across open moorland that is renowned for being dangerously exposed to the elements. Ice and snow are the enemies of the traveller setting out on this route to buy fresh milk, and we were determined that before the onset of winter proper we would solve the problem by getting a house cow.

It felt like a very natural, symbiotic relationship, to keep a cow to provide milk for the family on a daily basis. We have a small herd of pedigree beef Shorthorns, around twenty cows and a bull. Each year the cows produce a calf that we either retain to replenish the herd or sell on at around nine months of age as store cattle to other farmers who will fatten or breed from them. The Beef Short-horns, as the name suggests, have been specially

People ask if we bought her to save money on milk, but at £1,300 that is a lot of pints of milk before you break even!

bred to produce beef, and are a stockier version of the breed which originally had been dual purpose, bred for milk and meat production. That is not to say that the Beef Shorthorn does not produce milk – to be able to feed and nurture a calf they must lactate – but the cow is not going to milk the same volume as that of a dairy variety.

Our cows are in the main part docile (apart from Margaret) and there were occasions when we had been able to share the milk produced with a calf. It's a simple enough concept, and one that traditionally was commonplace on many farms. The freshly calved cow would feed her newborn calf for three or four days, the calf receiving the vital first milk, colostrum, that provides the energy and antibodies required. Then the calf is moved to a pen away from the cow overnight. The following morning the cow is milked, with just enough milk being taken for use in the household – then the calf is returned to its mother for the day. The cow's milk is shared between humans and calf, and if on any one day no milk is required in the household then the calf gets it all and there's no need to milk the cow.

This worked well for a while but, of course, as a larger-than-average household we consume a

decent quantity of milk on a daily basis, so we decided that we needed to buy a cow especially for this job. A dairy Shorthorn would fit the bill nicely and, with a breeder nearby in Kirkby Stephen, it was easy enough to go and buy a very pretty, delicate cow that was freshly calved and yielding seventeen litres of milk a day. People ask if we bought her to save money on milk, but at £1,300 that is a lot of pints of milk before you break even!

It has to be said that I didn't care for her name, and hoped that Wild Eyes did not reflect her nature. She was of a nervous disposition for a while – hardly surprising when you are taken to a new and unfamiliar home – but she soon settled in and was christened with the far gentler name of Buttercup.

The children enjoy helping out at milking time. Twice daily they will come along and fill her feed trough with cow ration, giving Buttercup a positive association of food with milking. As a result she is ready and waiting at the barn door every morning and night.

Once she settled into the daily routine, she yielded so much milk that we were struggling to use it all. As we didn't have her calf, which had been kept by her breeder, we bought a couple more, Bob and Stephen, to rear on all the excess milk. Next year when Buttercup calves we will be able to adopt the original idea of letting her keep and rear her own calf too, so this will save the milk glut.

The raw milk for the house is put through a stainless-steel sile with a dairy pad that filters out

any bits that might have contaminated the milk – maybe just a few wisps of hay – but other than that there is nothing else to do other than bottle it and cool it quickly. With a decent butterfat content the milk is creamy, and excellent in rice puddings and custards. There is something wholly satisfying about scrambled eggs for breakfast, the eggs coming from the chickens scratching around outside the back door coupled with a splash of milk fresh from the cow that morning. It is an absolute privilege to be able to indulge in such simple pleasures.

One day Violet informed me that she wanted to make cheese from our milk. I persuaded her gently that butter might be easier, as we already had a small glass churn that holds a couple of litres of milk. I was happy to see it finally being put into action, as it was one of those objects that had sat on the shelf for years gathering dust. We separated the cream from the milk by hand, using a slotted milk skimmer, and then filled the little churn. It is time consuming, especially when working with fresh cream – as Violet soon discovered. When her arms began aching, she convinced the others to take a turn. Finally her determination paid off and a solid lump of yellow butter stuck to the paddle in the blueish-tinged, fat-free buttermilk. The liquid by-product of butter-making is excellent as a milk replacement in scones or pancakes, so no element is wasted.

Chunky Beef Chilli

Prep time 10 minutes / Cooking time 2 hours / Serves 4

INGREDIENTS

2 tbsp Yorkshire rapeseed oil

400g boned shoulder beef,
  diced

1 medium onion, diced

1 tbsp ground cumin

2 tsp ground coriander

¼ tsp chilli powder

2 tbsp tomato puree

2 cloves garlic, crushed or grated

1 green chilli, halved and
  deseeded

1 cinnamon stick

400g can chopped tomatoes

300ml hot beef stock

480g jar roasted peppers,
  drained and roughly chopped

400g can kidney beans, drained
  and rinsed

30g dark chocolate, grated (70%
  cocoa solids)

1 small bunch coriander, roughly
  chopped, to garnish

To serve

150ml soured cream

1 lime, zested and cut into
  wedges

METHOD

1. Preheat the oven to 180°C/160°C fan/gas 4.

2. Heat 1 tablespoon of the oil in a heatproof casserole dish, add the beef, brown and seal the meat for 5 minutes, then remove from the pan to a bowl with a slotted spoon.

3. Return the pan to a medium heat with the remaining oil, add the onion and continue to cook for 5–6 minutes, until the onion has softened.

4. Return the meat and juices back into the pan, sprinkle over the dry spices and cook for 1 minute, until fragrant. Stir in the tomato puree, garlic, chilli halves and cinnamon, and mix well for 30 seconds.

5. Pour in the chopped tomatoes and stock, and bring to a simmer.

6. Cover with a lid, transfer to the middle shelf of the oven and cook for 1 hour 30 minutes until the meat is tender. Remove the lid, add the roasted peppers and kidney beans and cook for a further 15 minutes.

7. While the chilli cooks, mix the soured cream in a small bowl with the lime zest and season to taste.

8. Remove the chilli from the oven and allow to cool slightly, then stir through the chocolate and garnish with the coriander. Serve with rice, a dollop of zesty lime sour cream, and lime wedges.

Tip

Try serving with some wild rice and Mexican sides, such as homemade guacamole and tomato salsa.

Chocolate Rice Pudding with Pistachios

I love a rice pudding anyway, but the chocolate and pistachios just move it to another level. It is such a simple and filling way to warm up on a winter's day.

Prep time 5 minutes / Cooking time 40 minutes / Serves 4

INGREDIENTS

10g butter

100g pudding rice

40g golden caster sugar

50g dark chocolate, roughly
    chopped

2 tbsp cocoa powder

3 tbsp boiling water

700ml whole milk

40g shelled pistachio nuts,
    roughly chopped

METHOD

1. Preheat the oven to 160°C/140°C fan/gas 3.

2. Lightly grease a 22cm round deep enamel dish or baking dish with butter. Wash and drain the rice in a sieve, then pop into the dish. Sprinkle over the sugar and chocolate.

3. Mix the cocoa powder in a jug with the boiling water to form a paste then add to the dish with the rice and mix well to coat the rice in the cocoa. Pour the milk over the rice and give everything a good stir.

4. Put the dish onto a baking tray on the middle shelf of the oven for 40 minutes, until the rice pudding has thickened and is creamy. Stir the rice pudding a few times during cooking, as you do not want a skin to form.

5. Allow the pudding to cool slightly, then serve in bowls topped with pistachio nuts.

Tip
For an extra creamy pudding, add a swirl of double cream before serving.

December

A white Christmas is not so much of an unusual occurrence in the highest reaches of Swaledale. Nevertheless, here at Ravenseat, one of Yorkshire's remotest outposts, we still get excited at the prospect of waking up on Christmas morning to a snowy scene. We can have a blast of winter at any time from late October, but normally the first frozen, streaked smatterings of white appear on the bleakest exposed hillsides in November. Now the snow can fall heavily, no longer a meagre and short-lived dusting. As we look out towards the moortops from the comparatively sheltered basin in which the farmstead sits we can watch the weak milky winter light fade as it's engulfed by greying clouds, snowstorms rolling in from the east.

The quietness of the snow, the stillness and perfection in that early morning light, belies the trouble that it will inevitably cause, and that is why snow at Christmas time is such a delight. Only now do we get respite from the exterior pressures that hover on the periphery. Already the world will have slowed in readiness for the festive period; no need to worry about travel as the children are on their school holidays. The walkers

Ravenseat's east-facing arch window allows light in to waken the household at daybreak.

and visitors have gone home and the preparations that we have put in place mean that we have no need to go anywhere. Our acceptance of being snowed in at Christmas, with no one to please except ourselves, is profoundly uplifting. Weirdly, there is a sense of freedom from our entrapment. The work continues, but that's the joy – the focus, in fact. We are ready to wrap ourselves up in the simple tasks of daily life.

A proper winter, well there's a thing! Now, at the first sign of an Arctic blast or a sub-zero front sweeping the country, we are thrown into turmoil, shocked that Mother Nature should lash out. 'Nivver call out winter,' said old Jimmy, my friend and fellow Dales hill farmer who had lived through some of the most notable blizzards and storms in recent history without the amenities that we are so reliant upon. He meant, in his own unique manner, that in winter one should expect the bitter wind to chill you to the bone, be ready for leaden skies and drifting snow. We can rail against its cruelties, be forced to battle through these hard times, but we also trust that the land in its dormant state will, when timely, reawaken.

So rather than worry about a bout of wintery weather we should fret more over unseasonable

weather patterns. There was the time the children were swimming in the beck in late February on the day that we were pregnancy ultrasound scanning the yows when most years Adrian, our scanning man, is sat beneath a tarpaulin tent with only a flask of coffee and a cigarette staving off the numbing cold. There were hailstones in late June that battered and flattened our standing grass in the hay meadows, and the drought in spring that left us without water in our wettest pastures and led to us carrying buckets of water out into fields in which a beck flows that has never before run dry in our lifetime.

Being Swaledales, a native breed, the sheep are accustomed to coping with all weathers. They are adaptable, but we want them to thrive rather than just survive, so we must go that extra mile to see that they are well nourished during leaner times. We spare no effort in tending to the flock, leading them to shelter, and to kinder, fresher grazing and water if need be. There's an assumption that a sheep will never go thirsty when it snows, but a yow will sicken, and quickly too, if she doesn't have access to a puddle at least. Snow does not slake the thirst or have the same bodily effect as water, so we break the ice on a dew pond or resort to digging down through the snow to uncover a gutter still running with water. Snow has wonderful insulating properties that can either work in your favour or otherwise. If the ground beneath

the layer of snow was already frozen prior to snow-fall then it will be remain frozen, hard as iron, to an impossible depth, making it impenetrable.

After days spent battling in these conditions there's a newfound appreciation to be found in the comfort of the hearth with its stash of dry logs and coal, a kettle boiling on the hotplate and a verit-able inferno in the grate. On the fender, sodden waterproof leggings are laid steaming, while coats hanging from the flake drip from above. On the flagged floor, pools of water are to be sidestepped if a wet sock is to be avoided. Two terriers and mismatched gloves and hats are all vying for the position nearest to the hot surface of the warming oven. It is a yuletide scene fit for any Christmas card – though, rather than the scents of Nordic Fir and Spruce, we have the less-than-aromatic odour of damp dog fur and a room that could do with a jolly good spruce-up.

The horses are stabled over winter, outside most days and then inside for the night. This makes them more accessible (they can wander for miles when at the moor in summer should the fancy choose them) and a more acceptable form of transport (in Clive's eyes anyway, as it doesn't require an epic walk just to go and find them). During snowy times the horses are as good as anything mechanical, having an inbuilt sensor that seems to guide them around the drifts rather than through them, and it is rare that they should take a wrong turn. They seem to have a feel for where is safe to tread. Clemmy is very fond of getting out on the horses, and there is no weather that will curb her enthusiasm. She will not be dissuaded. Her supersonic hearing can detect the sound of a bridle being taken from the hook no

Raven rides out on Josie; it is one of the things she misses when she is away from home.

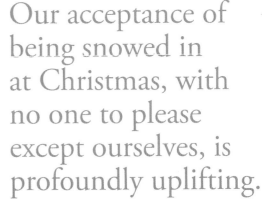

Our acceptance of being snowed in at Christmas, with no one to please except ourselves, is profoundly uplifting.

matter how far away she is, and the metallic rattle of the snaffle brings her running. 'Can I come?'

Tony's stumpy little legs struggle to keep up, having to do twice as many steps, cantering to keep up with Princess's trot. Tony and Clemmy follow in Princess's and my wake, a path cut through the snow that keeps him on track. No horseshoes are needed now; barefoot is better and prevents any ice balling up under their hooves, though clusters of snow cling to their feathers. And so we set off wearing plenty of layers of warm clothing, carrying a hip flask with something warming, and with a steadfast determination that we will not turn around at any cost.

All of the children have gone through stages of being pony mad. With some the passion for all things equine has remained; with others it has diminished somewhat, or at least been replaced with a love of horsepower of the other mechanical variety. Little Joe was our beloved Shetland who took them on adventures, taught them patience,

Even at Christmas everything needs feeding. The tutu seems to be a staple wardrobe item that has passed down from girl to girl – it has now gone from Clemmy, wearing it here, to Nancy!

to understand and listen and to trust. He also taught them how to sit out a flying change (in direction, rather than the type you'd find in a dressage test), how to be bold and fearless, and how to do an emergency dismount when he got the sudden urge to roll. He was a character of indomitable spirit right to the end. We lost him in the spring. After a hearty breakfast, a dose of treacle (as was his liking) and half a manger of our finest hay, he dropped dead. He was forty years old, but even though he was of such a great age we were not in any way ready for him to leave us. Little Joe was found lying motionless on a bed of deep straw in the stable with Tony standing over him, head bowed, bottom lip loose and quivering. I touched Joe. A procession of all the children, Reuben included, trooped in and stroked him tenderly. Dust and flecks of scurfy grey rose in the ray of bright sunshine that briefly illuminated the stable, Little Joe's body and the coat that once shone black as night. Tony snorted.

Not a word was said as Nancy gave him a gentle caress on his neck, still warm under his tangled, thick wiry mane. The straw beneath him had not moved; it was as though he had been struck down in full flight, even his tail, thinning through age, laid outstretched behind him as though he was at full gallop.

Reuben dug the grave and Little Joe was duly buried in the prominent and commanding position that he often frequented, before, as Raven succinctly put it, 'Pissin' off up't hill where we couldn't catch 'im.'

Hungry times are upon us now and, coupled with the celebratory aspect of Christmas and the New

Year, there's an overriding pressure to stock the dairy with all manner of seasonal festive produce that would not normally grace our table. There's a grave danger that I might panic, succumb to temptation and spend far too much money, time and effort in cooking something that realistically is not what we enjoy. It takes a certain amount of conviction to not conform, to resist being swayed into following a culinary trend. For a good many years I have served cold cuts of turkey and stuffing with homemade twice-cooked chips in gravy and the feedback has always been that it is THE best meal of all. It's hearty food, filling, warming and delicious.

Last Christmas came with its own unique issues, the lack of a poultry sale at the local auction mart being one of them. Every year we buy a turkey of epic proportions from one of the local farmers. The restrictions on numbers of people allowed to be together at Christmas due to the pandemic meant that large oversized turkeys were scarce, so rather than pay over the odds for a bird

of dubious origin we plucked up the courage (yes I know what I did there), to slaughter, ploat and pluck some of our own cock chickens. Sensitive to the fact that Miles tends his chickens with unerring devotion, we introduced the idea of eating a few of his flock very carefully. He was unexpectedly pragmatic in his approach to this request, and rather than being upset at the notion of dining upon his birds he saw it as a privilege to be providing the Christmas centrepiece.

We are all in good spirits, literally, with a bottle of table wine, some brandy for the Christmas pudding and sherry in the trifle that Raven made – and she was heavy-handed with the liquor. The chatter turns to Mary Moore, born in 1723, daughter of Christopher Spence. Mary died in the public house at Ravenseat in 1787 aged sixty-four! The children want to know what has changed since Mary keeled over in the bar beneath the optics. Clive is still lost in the moment, likely thinking of busty wenches and flagons of foaming ale.

Predictably the conversation is lively, energetic and occasionally ripe. As we make merry we consider what we know of the folks who have in the past lived out their lives right here, farming, entertaining guests and raising their families. Just how many generations have sat together around the kitchen table as we do? We will never really know; we have only just touched upon some elements of their stories, but we understand that this special and unique place lives on. It's a place of contradictions, homely yet inhospitable, a place entrenched in the past that now exemplifies our newfound modern values of caring for nature and the environment. It is, and always will be, as inspiring, invigorating and infuriating as it is incredible, such is the unshakeable and enduring nature of Ravenseat.

Roast Christmas Cockerel with Herb and Lemon Butter

Prep time 20 minutes, plus resting / Cooking time 2 hours 30 minutes / Serves 4–6

INGREDIENTS

3kg free-range cockerel

6g each of fresh thyme,
    rosemary and sage, leaves
    finely chopped

75g butter

1 lemon, zested and cut in half

250g streaky bacon

1 tbsp Yorkshire rapeseed oil

salt and black pepper

Gravy

30ml Marsala

25g plain flour

500ml hot chicken stock

METHOD

1. Remove the cockerel from the fridge 45 minutes before cooking. Preheat the oven to 200°C/180°C fan/gas 6.

2. Mix together the chopped herbs, butter and lemon zest in a small bowl. Push three-quarters of the butter under the cockerel skin, pulling the skin gently to get the butter right between the flesh and skin. Spread the remainder of the butter on top of the bird and season with salt and pepper.

3. Criss-cross the bacon over the bird and pop the lemon halves into the neck cavity. Drizzle over the oil, then place the bird into a large roasting pan. Roast on the bottom shelf of the oven for 30 minutes, then reduce the heat to 140°C/120°C fan/gas 1 and cook for a further 2 hours. Check that the meat juices run clear before removing the cockerel from the oven. If the cockerel is browning too much, place foil lightly over the top.

4. Once cooked, remove the cockerel from the oven and place on a tray. Cover with foil to rest for 30 minutes.

5. To make the gravy, strain the fat from the roasting pan, leaving only the juices. Heat the roasting pan, or add the juices to a saucepan, then pour over the Marsala and cook over a moderate heat, scraping the residue from the bottom of the pan. Take the pan off the heat and sprinkle over the flour; mix well to blend the flour with the meat juices. Cook over a gentle heat for 2–3 minutes, stirring, until the flour roux turns golden brown and it has a smooth texture.

6. Remove from the heat and stir in the stock then return to the heat and boil for 3–4 minutes, stirring all the time to remove any lumps and until the gravy thickens. Season to taste.

7. Carve the cockerel and serve with the gravy, roast potatoes and seasonal vegetables.

Tip
Save the cockerel carcass and use to make a homemade stock.

Christmas Cockerel Broth with Stilton Dumplings

Prep time 15 minutes / Cooking time 3 hours / Serves 4

INGREDIENTS

Stock

3 bay leaves

3 sprigs parsley

3 sprigs thyme

1 cockerel carcass, cleaned of skin, fat and meat

1 large onion, peeled and quartered

2 carrots

2 sticks celery

2 cloves garlic, crushed

5 peppercorns

Broth

1 tbsp Yorkshire rapeseed oil

1 medium leek, sliced

2 carrots, peeled and diced

250g swede, peeled and diced

200g potatoes, peeled and diced

200g soup mix, rinsed, drained and soaked overnight

1.5 litres fresh stock

300g shredded cooked cockerel

1 small bunch parsley, finely chopped

Dumplings

100g self-raising flour, plus extra for dusting

50g beef suet

50g Stilton, crumbled

2 tbsp parsley, chopped

salt and black pepper

METHOD

1. To make the stock, tie the fresh herbs together with some string to form a bouquet garni, then place all of the stock ingredients into a large pan and cover with cold water. Bring the pan to the boil, then cover and simmer gently for 2 hours. Strain the stock through a fine sieve into a large bowl and discard the carcass and vegetables. Allow to cool slightly, then skim off any excess fat.

2. To prepare the broth, heat the oil in a large ovenproof saucepan, add the leeks, carrots, swede and potatoes, and cook gently for 8 minutes to soften. Add the drained soup mix and pour over the fresh stock.

3. Bring to the boil and simmer with the lid on for 50 minutes, until the pulses have softened. Season to taste.

4. Meanwhile, while the broth cooks, make the dumplings. Preheat the oven to 190°C/170°C fan/gas 5. Mix the flour, suet, Stilton, parsley and seasoning in a bowl and gradually add 5 tablespoons of cold water to make a soft dough. Shape into eight balls on a lightly floured surface.

5. Stir the shredded cooked cockerel into the pan, then gently lower in the dumplings. Transfer the pan to the bottom shelf of the oven and cook for 18–20 minutes, until the dumplings are golden and crisp. Sprinkle over the parsley before serving.

Tip

The dumplings can be cooked for 20 minutes on the hob but if you prefer to cook them this way leave out the Stilton, as the addition of cheese does make the dumplings very soft.

Acknowledgements

There are so many people that deserve a mention; first and foremost my family, who of course feature heavily in this book – indeed, if I am in the picture then it is one of the children behind the lens.

Huge thanks to Jo and Jonathan Cantello, to Emma Donnan, and to everyone at Pan Macmillan, especially my publisher Ingrid Connell for giving me the opportunity to write this book, and for her absolute unerring patience at dealing with my hapless timekeeping. Thanks to Rebecca Needes, Lindsay Nash, Simon Rhodes, Holly Sheldrake and Siân Chilvers. And thanks to Heather Bowen for the lovely page design, and Jill Weatherburn for turning my notes into delicious recipes with proper measurements! To everyone else who worked tirelessly behind the scenes to help me along and support me, you know who you are.

Also by Amanda Owen

The Yorkshire Shepherdess
A Year in the Life of the Yorkshire Shepherdess
Adventures of the Yorkshire Shepherdess
Tales from the Farm

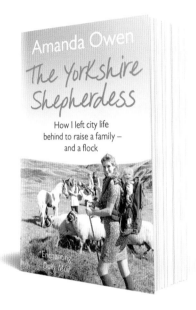

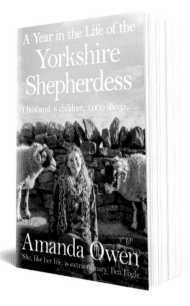

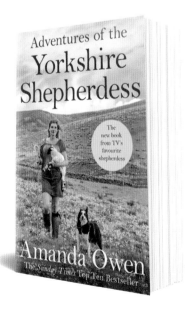

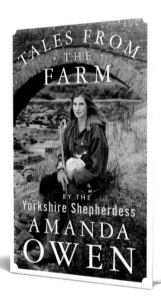

First published 2021 by Macmillan
an imprint of Pan Macmillan
The Smithson, 6 Briset Street, London EC1M 5NR
EU representative: Macmillan Publishers Ireland Ltd, 1st Floor, The Liffey
Trust Centre, 117–126 Sheriff Street Upper, Dublin 1, D01 YC43
Associated companies throughout the world
www.panmacmillan.com

ISBN 978-1-5290-5685-3

9 8 7 6 5

A CIP catalogue record for this book is available from the British Library.

Design and typesetting by Heather Bowen
Photographs © Amanda Owen
Recipe photographs © Jill Weatherburn
Illustrations © Jill Tytherleigh 2021

Printed and bound in Great Britain by Bell and Bain Ltd, Glasgow

FSC
www.fsc.org
MIX
Paper from
responsible sources
FSC® C116313

Visit **www.panmacmillan.com** to read more about all our books and to buy them. You will
also find features, author interviews and news of any author events, and you can sign up for
e-newsletters so that you're always first to hear about our new releases.